T0269373

Astronomers' Universe

More information about this series at http://www.springer.com/series/6960

Neil English

Space Telescopes

Capturing the Rays of the Electromagnetic
Spectrum

 Springer

Neil English
Fintry by Glasgow, United Kingdom

ISSN 1614-659X   ISSN 2197-6651   (electronic)
Astronomers' Universe
ISBN 978-3-319-27812-4   ISBN 978-3-319-27814-8   (eBook)
DOI 10.1007/978-3-319-27814-8

Library of Congress Control Number: 2016958306

Printed on acid-free paper

This Springer imprint is published by Springer Nature
The registered company is Springer International Publishing AG Switzerland
The registered company address is: Gewerbestrasse 11, 6330 Cham, Switzerland

Preface

Since the dawn of humankind, our species has sought to understand the nature of the universe around us and our place within it. For the vast majority of our history, we only had our immediate senses— sight, hearing, touch, taste, and a very limited sense of smell. On dark clear nights, huddled around a campfire, our ancestors gazed into the night sky, trying to make sense of the ever-changing view. Like the Sun and the Moon, the stars seemed to rise in the east and set in the west. Our ancestors noted the varying degrees of brightness of the stars in the sky, as well as their various hues—some red or orange, some yellow, and others white. And by carefully watching a few of the brighter luminaries, the most attentive sky gazers noticed that some of these stars wandered in their positions across the sky, sometimes disappearing altogether for months or years on end. These were called "planets," from the Greek word for wandering star, "planetos."

The invention of the telescope in Renaissance Europe ushered in a revolution in our understanding of the heavens that continues apace even today. Over four centuries, ground-based telescopes grew ever bigger and more powerful, but they could only see the universe in two principal wavebands—at visible light and at radio wavelengths. But advances in science were beginning to show that there existed other types of radiation—first infrared and then ultraviolet and more recently microwaves, X-rays, and gamma rays—all types of waves that are known collectively as electromagnetic radiation.

In order to see the universe at wavebands beyond the visible and radio parts of the electromagnetic spectrum, scientists had to find ways of penetrating the atmosphere, first by placing detectors on top of very high mountains where the air is thinner and then by using high-altitude balloons and sounding rockets, before the advent of the Space Age, when astronomers could finally build dedicated space-based missions with telescopes sensitive to the various parts of the electromagnetic spectrum blocked by our life-giving air. This is the story of that journey and the incredible new insights that transformed our knowledge of the cosmos utterly and forever. As well as being a story of human courage and single-mindedness, it also tracks the huge advances in technology humankind has enjoyed over the last century. In doing so, these space-based observatories have unveiled objects completely invisible to ground-based telescopes, demonstrating at once the beauty and the extreme danger of the space environment.

The book begins by taking a look at the science of waves, particularly electromagnetic waves, and their behavior and detection. In this opening chapter, we explore the source of much of the electromagnetic radiation in the cosmos—atoms. We'll survey the origin and nature of spectra and how this knowledge helps astronomers unravel a veritable treasure trove of new information about the chemical constitution and physical conditions of the stars and other astrophysical bodies generating them. Finally, we cover some basic astronomy that will help us fully engage with the science discussed in later chapters.

Chapter 2 recounts the incredible allegory of the Hubble Space Telescope, its perilous early days when engineers discovered its giant 96-inch mirror was misshapen, followed by its correction by NASA astronauts. We then explore the rich heritage of images captured by the world's most famous space telescope and how it completely transformed our understanding of the universe, both nearby and billions of light-years away.

We continue our exploration of the electromagnetic spectrum by chronicling the development of infrared (IR) astronomy and how space-based IR telescopes have allowed us to peer deep inside dust-laden star clusters and galaxies, hunting down a variety of cooler celestial objects invisible to even the largest optical telescopes. This chapter discusses important IR telescopes, including ROSAT and the Spitzer space telescope.

Next, we recount the fascinating story of the high-energy universe, beginning by exploring the shortest electromagnetic waves of all—gamma rays—and the extraordinary history of how scientists and engineers built better gamma ray detection systems, carrying them aloft on sounding rockets, as well as a host of sophisticated gamma ray space telescopes over many decades. These include early satellites such as Cos-B, Compton, BATSE, HETE, and more high-tech spacecraft including BeppoSAX, INTE-

GRAL, Swift, and Fermi. Here, we shall explore the mysterious nature of some of the most violent explosions in the universe: gamma ray bursts.

In the next chapter, we explore a waveband closer to the visible region of the electromagnetic spectrum—the ultraviolet universe. While a trickle of long-wave ultraviolet (UV) radiation can penetrate Earth's atmosphere, the vast majority of this radiation can only be detected in space. Space-based UV astronomy got a great boost in the 1970s with the launch of TD-1A and the Dutch ANS satellite. With the advance of technology, more sophisticated UV observatories came to the fore, with the International Ultraviolet Explorer (IUE) and the American-led Extreme Ultraviolet Explorer (EUVE), allowing astronomers to make great leaps forward in understanding a host of astrophysical phenomena from active galaxies, massive young stars, and even new insights into solar system objects.

X-rays have long provided us with a means of seeing the invisible, but it was not for several decades after their discovery that astronomers began to view the cosmos at these wavelengths. Of all the wavebands of the electromagnetic spectrum, it is arguably X-rays that have revealed the most insight into the physics of the Sun and hot OB and A stars. Astronomers have also discovered that cool, M dwarfs emit prodigious amounts of X-ray flares, calling into question whether the planets they harbor could ever sustain life. The earliest X-ray detectors were very primitive, but over the decades, they became increasingly more powerful. Accordingly, we shall explore key X-ray astronomical observatories, including the Orbiting Astronomical Observatory satellites, followed by more specialized missions, including Copernicus, Einstein, Uhuru, and Chandra.

Next, we return to wavelengths that are far too long to be seen by the human eye. The birth of microwave astronomy was essentially ground based, when in 1964 Arno Penzias and Robert Woodrow Wilson discovered the cosmic microwave background radiating almost uniformly across the entire sky. The significance of this serendipitous discovery cannot be overstated, since the unveiling has provided brand-new insights into how the universe must have begun. Described by Robert Gamow as the afterglow of creation, this radiation represents the remnants of the primordial fireball that characterized the hot Big Bang universe. The momentous discovery was followed up by the highly ambitious space missions COBE and, more recently, WMAP. These observatories gained glimpses into how galaxies and their constituent stars emerged from cosmic chaos and how their findings revolutionized our ideas about the origin and evolution of our universe.

By observing the universe across many wavebands, astronomers can gain a complete picture of the underlying nature of astrophysical bodies. It is arguably the Sun that has benefitted most from studies across the electromagnetic spectrum, and, in this capacity, we devote an entire chapter to how this knowledge was applied to our life-giving star and how the contributions made at visual, UV, X-ray, and gamma ray wavelengths have enabled us to piece together our most detailed picture yet of our life-sustaining Sun, with its dark spots, plages, flares, prominences, and much more. The collective data from across the EM spectrum has provided a more complex picture into the nature of the Sun and, by implication, other stars that inhabit the universe.

The era of precision astrometry, that is, the science of measuring the vast distances to the stars, entered a new era with the launch of the Hipparcos satellite, which cataloged 118,200 stars during a 4-year mission between 1989 and 1993. The spectacular success of Hipparcos was added to with the launch of the European Space Agency's Gaia spacecraft, the ongoing mission of which is to record the positions of up to one billion objects in the heavens in 3-D. In the last decade or so, astronomers have begun to employ orbiting satellites to hunt down and characterize a plethora of extrasolar planets, that is, planets orbiting other stars. We explore the recent success of the Kepler planet-finding mission, how it detects these planets, and the results of the surveys so far.

Finally, we end not with history but with the near future, by taking an in-depth look at the greatly anticipated replacement for the Hubble Space Telescope—the James Webb Space Telescope—which is widely expected to boldly go where no telescope has gone before.

Above all, however, the story of the history of space telescopes is one of human ingenuity, single-mindedness, collaboration, setbacks, and success—a microcosm of the human condition itself.

Fintry by Glasgow, UK   Neil English

Acknowledgments

I would like to thank John Watson and Maury Solomon at Springer for endorsing this project through its various stages of development. I would also like to extend my thanks to Michael Carroll for excellent copyediting of the raw text. Last, but certainly not least, I want to thank my wife, Lorna, and my two sons, Oscar and Douglas, for putting up with my many hours away from them in order to get the writing work done for this manuscript.

Contents

1. Light: Nature's Great Information Superhighway

Humanity's curiosity about the cosmos stems from our ability to see. On a clear, dark night, our eyes can make out a few thousand bright pinpoints in the sky, stars located at great distances from Earth.

In this chapter, we shall explore the basic nature of light as understood by physicists, and explore how visible light is just one narrow window within a much broader phenomenon known as electromagnetic radiation.

Diligent investigation carried out over the centuries has established that light exhibits wavelike properties. Waves are oscillatory phenomena that carry energy and momentum through matter or empty space. All waves have a wavelength, that is, a fixed distance between successive crests or troughs. Wavelength is measured in meters. The number of whole waves passing a reference point in one second defines its frequency, measured in units called Hertz. Finally, all waves move through a medium with a certain velocity. This velocity is determined by multiplying the frequency by the wavelength. Light waves move at a speed of 300,000 km per second through the vacuum of space and at lower velocities through transparent materials (Fig. 1.1).

All waves exhibit a number of properties in common, including:

1. reflection
2. refraction
3. diffraction
4. interference

Reflection The reflection of waves is something we are aware of in our everyday lives. We look in the mirror and see our image. Sometimes we hear sound echoes. Both of these are examples of

© Springer International Publishing Switzerland 2017
N. English, *Space Telescopes*, Astronomers' Universe,
DOI 10.1007/978-3-319-27814-8_1

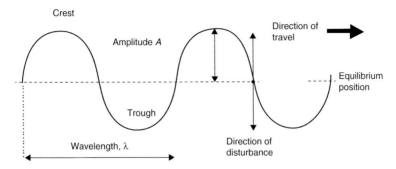

FIG. **1.1** The basic properties of a wave (Image by the author)

reflection. The reflection of waves makes possible the ability to bring waves to a precise focus, allowing us to see things that are quite invisible to the unaided eye. Reflection of waves enables great telescopes to see into the distant universe and permits our global technical civilization to exist via satellite communications.

Refraction When a wave enters a new medium such as glass or perspex, it slows down, and, unless it enters at right angles to the surface, it will change direction. This results in a bending of light that we call refraction. Refraction is the mechanism by which curved glass lenses can bring light to a precise focus. Different materials vary in their ability to bend light. Furthermore, a given transparent material bends different wavelengths of light by differing amounts. One way of quantifying this degree of bending is measured by a quantity known as the refractive index of the material. In general, the longer the wavelength, the lower the refractive index. The latter is responsible for the beautiful rainbow of colors seen when white light passes through a prism or a raindrop. Red light deviates least, and violet light, the most.

Diffraction When a wave passes an obstruction like a hill or through a gap, the wave spreads behind the regions that were obstructed. This is why radio waves can be detected behind buildings and hills. Such waves are said to be diffracted. If a wave is directed through a gap, it continues through that gap and spreads out into the area shielded by the gap. It follows that long waves are diffracted less than short waves, explaining why long radio waves can be picked up more easily inside a deep valley than shorter waves.

Interference Two waves with the same velocity, frequency and wavelength that hold their positions relative to each other are said to be coherent. Two kinds of interference are possible, constructive and destructive. When two waves meet each other in such a way that the peak of one wave meets a crest of the other, the waves 'add up' and create a locus of constructive interference. Conversely, when the crest of one wave lines up with a trough of another wave, they cancel each other out and exhibit so-called destructive interference. The interference of light is responsible for a number of everyday phenomena, including the pretty colors seen when oil floats on water or when a soap bubble floats in the air (Fig. 1.2).

Visible light is just one small part of what physicists refer to as the electromagnetic (EM) spectrum. Visible light consists of the red waves at the lowest frequency and violet light at the highest visible frequencies (Fig. 1.3).

Beyond the red light lie the invisible rays of infrared radiation. William Herschel is credited for having first discovered this radiation back in 1800, after demonstrating that a thermo meter placed just beyond the red end of the visible spectrum registered a temperature rise. Longer still are microwaves, which have an interesting story behind their discovery. While conducting research on radar, the American radar pioneer Percy Spencer found that a chocolate bar had melted in his shirt pocket. Further investigation

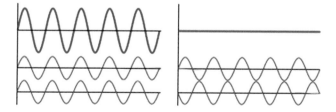

FIG. 1.2 Schematic showing constructive (*left*) and destructive (*right*) interference. The resultant effect is illustrated in the top waves (Image by the author)

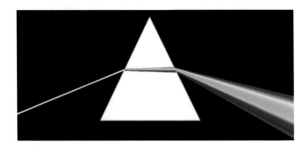

FIG. 1.3 When white light passes through a glass prism, a spectrum of colors is produced (Image by the author)

by Spencer showed that these 'heat rays' had wavelengths longer than infrared rays, and he called them microwaves. His experience inspired Spencer to invent the microwave oven.

Electromagnetic waves of even lower frequency are called radio and TV waves. Radio waves were first predicted by Scottish mathematical physicist James Clerk Maxwell in 1867. Maxwell noticed wavelike properties of light and similarities in electrical and magnetic responses. He then proposed equations that described light waves and radio waves as waves of electromagnetism that travel in space, radiated by a charged particle as it undergoes acceleration. In 1887, Heinrich Hertz demonstrated the reality of Maxwell's electromagnetic waves by experimentally generating radio waves in his laboratory. Many inventions followed, making use of these 'Hertzian' waves (another term for radio waves) to transfer energy and information through space.

If we venture beyond the violet, we encounter waves of ever increasing frequency from ultraviolet, X-rays and finally gamma rays. After hearing about Herschel's discovery of an invisible form of light beyond the red portion of the spectrum, the Polish scientist Johann Ritter decided to conduct experiments to determine if invisible light existed beyond the violet end of the spectrum as well. In 1801, he was experimenting with silver chloride, a chemical that turns black when it is exposed to sunlight. He had heard that exposure to blue light caused a greater reaction in silver chloride than exposure to red light. Ritter decided to measure the rate at which silver chloride reacted when exposed to the different colors of light. To do this, he directed sunlight through a glass prism to create a spectrum. He then placed silver chloride in each color of the spectrum. Ritter noticed that the silver chloride showed little change in the red part of the spectrum, but increasingly darkened at the violet end of the spectrum. This showed that exposure to blue light did cause silver chloride to turn black much more efficiently than exposure to red light.

Ritter then decided to place silver chloride in the area just beyond the violet end of the spectrum, in a region where no sunlight was visible. To his amazement, he saw that the silver chloride displayed an intense reaction well beyond the violet end of the spectrum, where no visible light could be seen. This showed for the first time that an invisible form of light existed beyond the violet end of the visible spectrum. This new radiation, which Ritter called 'chemical rays,' later became known as ultraviolet light or ultraviolet radiation; 'ultra' meaning beyond. Ritter's experiment, along with Herschel's discovery, proved that invisible forms of light existed beyond both ends of the visible spectrum.

In late 1895, a German physicist, W. C. Roentgen, was working with a cathode ray tube in his laboratory. He was working with tubes similar to contemporary fluorescent light bulbs. He evacuated the tube of as much air as possible, replaced it with a special gas, and then passed a high electric voltage through it. When he did this, the tube would produce a fluorescent glow. Roentgen shielded the tube with heavy black paper and found that a green-colored fluorescent light could be seen coming from a screen sitting a few feet away from the tube. He realized that he had produced a previously unknown "invisible light," or ray, that was being emitted from the tube—a ray that was capable of passing through the heavy paper covering the tube. Through additional experiments, Roentgen also found that these new rays would pass through most substances, casting shadows of solid objects on pieces of film. In mathematics, chemistry and other areas of study, "X" is used to indicate an unknown quantity, so he named the new rays X-rays.

Further experiments conducted by Roentgen found that X-rays would pass through living tissue, rendering bones and metals visible. One of Roentgen's first experiments late in 1895 was a film of his wife Bertha's hand with a ring on her finger. The news of Roentgen's discovery spread quickly throughout the world. Scientists everywhere could duplicate his experiment because the cathode tube was very well known during this period. By early 1896, X-rays were already being utilized clinically in the United States to investigate such things as bone fractures and other kinds of wounds.

The first gamma ray source to be discovered was through the radioactive decay of a substance. When a substance radioactively decays, an excited radioactive nucleus emits a gamma ray, and the resulting nucleus is more stable. Paul Villard, a French chemist and physicist, discovered gamma radiation in 1900 while studying radiation emitted from radium. Villard knew that his described radiation was more powerful than previously described types of rays from radium, which included beta rays, first noted as "radioactivity" by Henri Becquerel in 1896, and alpha rays, discovered as a less penetrating form of radiation by Rutherford in 1899. However, Villard did not recognize them as a fundamentally new type of electromagnetic radiation. This was accomplished by Ernest Rutherford, who in 1903 named Villard's rays "gamma rays" by analogy with the beta and alpha rays that Rutherford had differentiated in 1899. The "rays" emitted by radioactive elements were named in order of their power to penetrate various materials, using the first three letters of the Greek alphabet: alpha rays as the least penetrating, followed by beta rays, followed by gamma rays as the most penetrating. Rutherford also noted that gamma rays were not deflected (or at least, not easily deflected) by a magnetic field, another property distinguishing them from alpha and beta rays.

Although EM waves exhibit different frequencies (and hence wavelengths), they all move at the same velocity through a given material, as described by Maxwell's famous equations. Figure 1.4 shows the relative positions of these various forms of EM radiation.

In the dawning years of the twentieth century, scientists uncovered new phenomena about the nature of light that could best be explained if light and other forms of EM radiation acted like a particle and not a wave. Perhaps the best attested example of this is the photoelectric effect. When light of a given frequency is shone onto a polished metal surface, it develops a positive electric charge. This is due to the ejection of negatively charged electrons from the metal atoms near its surface. Moreover, only light with a frequency at or above the so-called threshold frequency can elicit the effect.

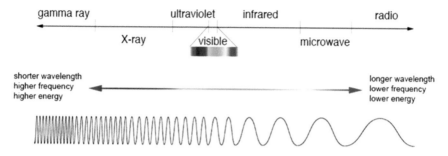

FIG. 1.4 The electromagnetic spectrum (Image by the author)

If the light that shines on the metal has a frequency below the threshold frequency, no photoelectric effect is seen, no matter how intense the light source or how long one irradiates the metal surface. These strange results can be interpreted as light behaving like tiny packages of energy or 'photons,' more like bullets than waves. Moreover, the energy of a photon is directly proportional to its frequency; the higher the frequency, the higher the energy of the photon. That's why visible light doesn't cause sunburn while higher energy (higher frequency) ultraviolet light does. It also explains why higher frequency X-rays and gamma rays are even more dangerous to living things.

German physicist Max Planck built on this work. Planck derived a law for the spectrum of black body radiation. A black body is an object that absorbs all the radiations that is incident upon it (it reflects no radiation) and radiates energy that is only from the object itself. An example of a black body is a furnace with walls that absorb all incident radiation but has a very small opening for radiation from inside the furnace to escape. Wilhelm Wien derived a simple formula that predicts the temperature of any such body from knowing its peak wavelength (Lambda) output. Specifically:

$$\text{Lambda Peak (in meters)} \times T = 2.889 \times 10^{-3}$$

Wien's displacement law allows astronomers to compute the temperature of any black body (of which stars approximate very well) if we know the wavelength at which the body emits its peak wavelength. In effect then, Wien's displacement law enables us to measure the temperature of any black body source, thereby acting as a cosmic thermo meter! Wien's law shows us that the hotter the body's temperature, the more it moves to shorter wavelengths (higher frequencies). For example, our Sun radiates its peak wavelength at 500 nm. Plugging this into Wien's formula and rearranging for T gives 5778 K. Cooler stars have their peak wavelengths shifted to the red end of the visible spectrum, while hot-white and blue-white stars display their peak wavelengths in the blue and violet part of the visible spectrum (Fig. 1.5).

Human ingenuity has enabled us to build instruments that can detect all forms of electromagnetic radiation, from ultra-short wave gamma rays through the visible spectrum and on through the infrared, microwave and radio regions of the EM spectrum. Infrared, microwaves and radio waves can be focused using curved parabolic surfaces. The larger the wavelength, the larger the size of the dish needed to collect and focus those waves. Ultraviolet radiation can also be focused by telescopes in the same way that they collect visible light.

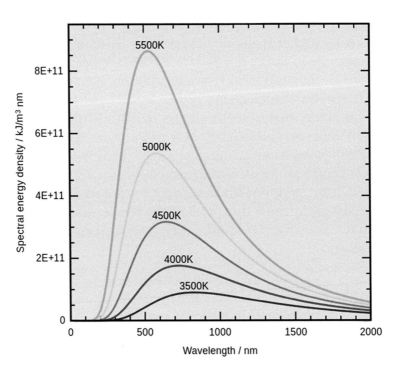

FIG. 1.5 Wien's law—the hotter the object the more its peak wavelength shifts to shorter wavelengths (Image by the author)

However, focusing shorter wave radiation—X-rays and gamma rays—requires a different kind of technology to that employed for other parts of the EM spectrum. These high energy photons cut right through regular mirrors. The rays must be bounced off a mirror at a very low angle if they are to be captured. This technique is referred to as grazing incidence. For this reason, the mirrors in X-ray telescopes are mounted with their surfaces only slightly off a parallel line with the incoming X-rays. Application of the grazing-incidence principle makes it possible to focus X-rays from a cosmic object into an image that can be recorded electronically (Fig. 1.6).

Several types of X-ray detectors have been devised, including Geiger counters, proportional counters, and scintillation counters. These detectors require a large collecting area because celestial X-ray sources are remote and so are almost invariably weak; thus a high efficiency for detecting X-rays over the cosmic ray-induced background radiation is required. Although astronomers using visible and UV telescopes typically adopt familiar concepts such as wavelength or frequency to describe the range of EM radiation that they are sensitive to, X-ray and gamma ray telescopes make use of another unit—the electron volt (eV). One electron volt is equivalent to 1.6×10^{-19} joules. In general, X-ray telescopes cover an energy range between 1 keV to 10 MeV, while their gamma ray counterparts monitor higher energy radiations up to several tens of GeVs. In addition, astronomers refer to

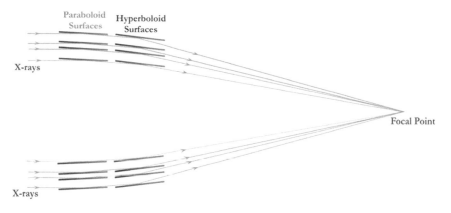

FIG. 1.6 X-ray and gamma ray telescopes focus rays by reflecting them off mirrors at very shallow angles (Image by the author)

higher energy radiations as 'hard,' while those wavebands with less energy are said to be 'soft.' On our journey through the history of space telescopes, we will occasionally encounter a less commonly used unit of energy called the erg. One erg is the equivalent of 10^{-7} joules. Thus, power output can also be expressed in units of ergs per second.

Now that we have briefly surveyed the properties of EM radiation, it is time to look at the kind of scientific work these telescopes can accomplish in the vacuum of space. Space is a hostile place. Earth's atmosphere provides a protective blanket to shield surface life forms from the deadly rays emanating from a variety of celestial sources. The atmosphere allows visible light and radio waves to pass through it, but blocks the majority of infrared radiation, ultraviolet, as well as deadly X-rays and gamma rays. For these reasons, the majority of space telescopes are dedicated to imaging these radiations.

Turbulence in Earth's atmosphere can distort the images produced by large optical telescopes, even those placed at a very high elevation. As a result, astronomers have sought to launch optical telescopes into the vacuum of space in search of the clearest images of the cosmos possible. The Hubble Space Telescope, discussed at length in a later chapter, is the most celebrated example from this genre. By imaging celestial objects in different regions of the EM spectrum, astronomers can glean far more information than they could if they restricted their studies to the visible wavelengths (Fig. 1.7).

As well as imaging objects both within and beyond our own Solar System, starlight reveals far more information when subjected to spectroscopic study. By passing light through a high-resolution diffraction grating, astronomers can unravel the chemical composition of the star with high precision. In addition to its bulk chemical constitution, the temperature, space velocity, magnetic field strength, and even the tell-tale signs of planets orbiting their parent stars can be studied. In the chapters that follow, we shall explore how telescopes launched into Earth orbit and beyond peer into the deep universe with eyes sensitive to different parts of the EM spectrum. The mind still boggles at what they have seen, as later chapters shall reveal.

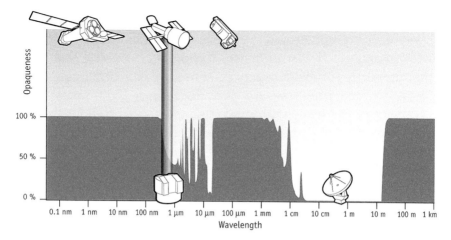

Fig. 1.7 This shows the opaqueness of Earth's atmosphere to various parts of the EM spectrum (Image credit: ESO)

Divining the Chemistry of the Stars

One of the best examples of false prognostications regarding the limits of science, in this case astrophysics, was made in 1835 by the prominent French philosopher Auguste Comte. In his *Cours de Philosophie Positive* he wrote:

> On the subject of stars, all investigations which are not ultimately reducible to simple visual observations are … necessarily denied to us. While we can conceive of the possibility of determining their shapes, their sizes, and their motions, we shall never be able by any means to study their chemical composition or their mineralogical structure … Our knowledge concerning their gaseous envelopes is necessarily limited to their existence, size … and refractive power, we shall not at all be able to determine their chemical composition or even their density… I regard any notion concerning the true mean temperature of the various stars as forever denied to us.

However, only 14 years later, the physicist Gustav Kirchhoff discovered that the temperature and chemical composition of a gas could be deduced from its electromagnetic spectrum viewed from an arbitrary distance. This method was extended to astronomical bodies by Huggins in 1864, using a spectrograph attached to a telescope. Not only have we learned how to determine the chemical composition of the stars and nebulae, but the element helium (the second most abundant in the universe) was first identified in the spectrum of the Sun, rather than in an earthbound laboratory.

Today, astronomers use spectroscopy to measure chemical abundances, temperatures, velocities, rotation rates, ionization states, magnetic fields, pressure, turbulence, density, and many other properties of distant planets, stars, and galaxies. Some of the objects studied this way are over 10 billion light years away. Spectroscopy is a study that yields rich information about the universe. And within a laboratory framework, spectroscopy provided the experimental basis for the most accurate science we know of: quantum mechanics.

When a body is heated, and it becomes more energetic, it will begin to emit visible light. In the early nineteenth century, physicists were aware that the light from the Sun was missing wavelengths characterized by dark lines in the resulting emission spectrum. These were called Fraunhofer lines after the German physicist and optician Joseph Von Fraunhofer, who discovered and recorded in excess of 500 of them.

Around the same time, chemists such as Robert Bunsen (among others) began examining the light emitted by chemical elements when heated in a laboratory flame. When this light was passed through a spectrometer to more closely analyze the spectrum produced, it showed that it was made up of a series of discrete lines. Each series was unique to a particular element and could thus be used as a type of 'signature' to determine the element's presence. The visible spectrum of the element hydrogen is shown in Fig. 1.8.

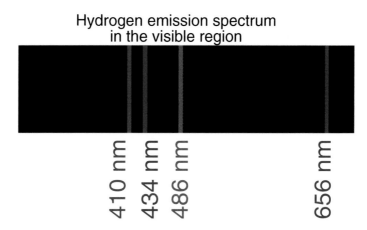

FIG. 1.8 The visible spectrum of hydrogen (Image credit: University of Texas)

When chemists unraveled more about the nature of the atom, they discovered that the more complex atoms also had more complex spectra. Illustrated below are the visible spectra of carbon, nitrogen, and oxygen. It became clear that the 'missing' lines in the Sun's continuous emission spectrum were matched with the emission lines of individual elements, which could be measured here on Earth. As a result, it was proposed that the white light being emitted by the surface of the Sun was passing through its atmosphere of chemical elements, which were only absorbing the wavelengths that somehow characterized their structure. This discovery led to many further questions. For one thing, why do the wavelengths generated by hydrogen differ from those absorbed by another chemical element such as iron or zinc? What's more, why is it that only these special wavelengths are being absorbed or emitted to the exclusion of all the others? The answers to these questions had to wait 60 more years to be addressed and required the work of great theoretical physicists, such as Max Planck, Albert Einstein, and Niels Bohr.

Einstein developed Planck's work on black body radiation and proposed that, at the smallest level, light was delivered in packets of energy called photons. Each particle carried a fixed amount of energy given by the equation $E = hf$, where E represents energy, h is Planck's constant, and f is the frequency of the light considered. This means that each photon of light absorbed or emitted by an atom has a particular, unique energy.

Bohr's model of the hydrogen atom uses this fact and Rutherford's nuclear atomic model to propose that electrons orbit a dense, positively charged nucleus in a manner similar to planets orbiting a star. The radius of the orbit is determined by the electrical potential energy of the electron in the electric field created by the nucleus. As the electrons are attracted to the nucleus, work needs to be done to move them further away and so larger radius orbits are higher-energy states. More significantly, only particular energy states are permitted. In other words, these orbits are quantized. This means that only certain, particular orbit radii can be maintained.

To jump from a low energy state to a high energy state requires the electron to absorb exactly the right amount of energy from an incoming photon. This is known as an electron excitation with the electron being promoted to a higher energy state. Conversely, if an electron falls from an excited state to a lower state, it will emit a photon whose energy will correspond exactly to an energy gap in the two states.

The proposal to restrict the energy states to a limited set of discrete values explains the observed absorption and emission spectra. The characteristic lines or wavelengths emitted or absorbed (missing) correspond exactly to the energy gap in the two electron states. As different elements contain different numbers of positively charged protons in the nucleus (and hence the number of electrons in the orbits), the exact values for the energy states and their differences are unique to those elements (Fig. 1.9).

The spectral lines are not restricted to visible wavelengths. Many other lines were uncovered in different parts of the EM spectrum. Thus, scientists could use visible, infrared, UV, X-ray, and even gamma ray spectroscopy to divine further secrets of the composition and physical properties of celestial objects.

Though it is beyond the scope of this text to discuss the full panoply of details that spectroscopy reveals about a luminous body, it is enough to say the study reveals an enormous amount of detail about the temperature, surface gravity, mass, ionization state of its constituent matter, motion (especially the Doppler effect, discussed briefly in the next section of this chapter), and magnetic field strength of the celestial body. Spectroscopy may even enable astronomers to sniff the chemistry of the atmospheres of planets orbiting other stars. We could then identify whether life could exist on these planets! We shall explore these details in later chapters while discussing specific space telescope missions (Fig. 1.10).

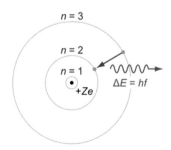

FIG. 1.9 The three lowest electron orbital states permitted in the Bohr model of the hydrogen atom (Image credit: Jabberwok)

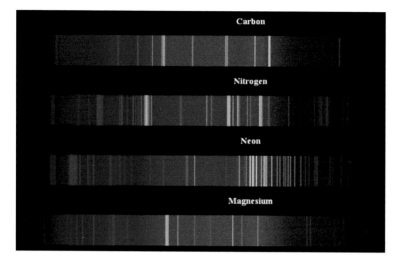

F<small>IG.</small> **1.10** The emission spectra in visible light for some common elements (Image credit: spiff.rit.edu)

Electromagnetic Waves and Motion

The history of this subject began with the development in the nineteenth century of the theory of wave mechanics and the exploration of phenomena associated with the Doppler effect. Imagine you are watching an ambulance approach you at high speed with its siren blasting. As the sound waves are moving with their own velocity, the waves are compressed in the direction of motion, causing their frequency (pitch) to increase. Conversely, as the ambulance recedes from you, the waves are stretched out and their pitch accordingly decreases. This effect was named after the Austrian scientist, Christian Doppler (1803–1853), who offered the first known physical explanation for the phenomenon in 1842. The hypothesis was tested and confirmed for sound waves by the Dutch scientist Christophorus Buys Ballot in 1845. Doppler correctly predicted that the phenomenon should apply to all waves, and in particular, suggested that the varying colors of stars could be attributed to their motion with respect to Earth. Specifically, pitch is to sound as color is to light. Before this was verified, however, it was found that stellar colors were primarily due to a star's temperature, not motion. Only later was Doppler vindicated by verified redshift observations.

The first Doppler red shift was described by French physicist Hippolyte Fizeau in 1848, who pointed to the shift in spectral lines seen in stars as being due to the Doppler effect. If the star is moving away from us, then the spectrum of Fraunhofer lines would be shifted to lower frequencies, that is, 'red-shifted.' Conversely, if the star is moving towards a stationary observer, those same spectral lines shift to higher frequencies, that is, 'blue-shifted.'

The physical quantity called red shift (z) is provided by the simple formula:

$$z = \left(\text{Lambda}\left(\text{observed}\right) - \text{Lambda}\left(\text{rest}\right)\right) / \text{Lambda}\left(\text{rest}\right)$$

Furthermore, for stars or galaxies moving at speeds much less than the speed of light, the velocity of the star (v) can be calculated using the formula:

$$v = zc$$

where c is the speed of light in a vacuum (300 million meters per second). The effect is sometimes called the Doppler-Fizeau effect.

In 1868, British astronomer William Huggins was the first to determine the velocity of a star moving away from Earth by this method. In 1871, optical red shift was confirmed when the phenomenon was observed in Fraunhofer lines using solar rotation, about 0.01 nm in the red. In 1887, Vogel and Scheiner discovered the annual Doppler effect, the yearly change in the Doppler shift of stars located near the ecliptic due to the orbital velocity of Earth. In 1901, Aristarkh Belopolsky verified optical redshift in the laboratory using a system of rotating mirrors.

The earliest mention of the term "red shift" in print (in hyphenated form) appears to be by American astronomer Walter S. Adams in 1908, when he mentions, "Two methods of investigating that nature of the nebular red-shift." The word does not appear unhyphenated until about 1934 by Willem de Sitter, perhaps indicating that up to that point its German equivalent, *Rotverschiebung*, was more commonly used.

In 1912, Vesto Slipher, then based at Lowell Observatory in Flagstaff, Arizona, discovered that most spiral galaxies, which were mostly thought to be spiral nebulae, had sizable red shifts. Slipher first documented his measurement in the very first volume of the

Lowell Observatory Bulletin. Just a few years later, he wrote a review in the journal *Popular Astronomy.* In it he states, "...the early discovery that the great Andromeda spiral had the quite exceptional velocity of –300 km/s showed the means then available, capable of investigating not only the spectra of the spirals but their velocities as well." Slipher reported the velocities for a total of 15 spiral nebulae from across the night sky, 12 of which having observable "positive" (that is, recessional) velocities. Just over a decade later, Edwin Hubble discovered an approximate relationship between the red shifts of such 'nebulae' and the distances to them with the formulation of the famous law that bears his name; Hubble's law. These observations vindicated Alexander Friedmann's 1922 work, in which he derived the groundbreaking equations still used by cosmologists today. Together, they are considered robust evidence for an expanding universe and the Big Bang theory.

The Zeeman Effect

As well as providing information about the chemistry and speed of astronomical objects, spectra can also be used to measure the intensity of magnetic fields in the body under study.

This can be achieved through a process called the Zeeman effect, which involves the splitting of a spectral line by a magnetic field. Any single atomic spectral line under normal conditions would be split into two in the presence of a magnetic field, one of lower energy and one of higher energy in comparison to the original line. The detailed explanation for the Zeeman effect requires some understanding of quantum physics, which need not concern us further here apart from stating that the strength of the magnetic field inside a star can be measured using this technique (Fig. 1.11).

A Brief Look at the Lives and Classifications of Stars

In astronomy, stellar classification involves grouping stars together based on their spectral characteristics. The spectral class of a star gives clues about the temperature of its atmosphere. Most stars

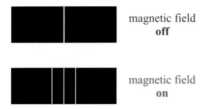

FIG. 1.11 The Zeeman effect occurs when spectral lines are split due to being subjected to a magnetic field (Image credit: www.periodni.com)

are currently classified using the letters O, B, A, F, G, K, M, and L, in order of increasing coolness. The hottest spectral classes (O, B, and A) have a blue-white or white color, those of intermediate spectral class (F and G) are yellow-white, K stars are orange, and M-type stars shine with a red hue. L-type stars, which are legion in the galaxy, shine most strongly, not in visible light, but in the infrared.

Our Sun, a G-type star, formed about 4.6 billion years ago and is now about half way through its life. Research has shown that during this epoch, the levels of the radioactive elements—uranium and thorium—essential for building planets and plate tectonics, were at their highest. Moreover, there is now evidence that the solar nebula—that great cloud of gas and dust out of which the Sun and its retinue of planets formed—were enriched by an eclectic mix of elements from not one but two types of stellar explosions (supernova events). These exploding stars were not so close as to destroy the solar nebula, but neither were they so far away as to make no difference to its final chemical constitution. How uncanny!

Scientists have discovered many different kinds of stars—from white dwarfs to supergiant stars. But up until the early 1900s, there was no general way to classify them. All of that changed with the invention of the Hertzsprung-Russell (H-R) diagram, which has become one of the most important tools in stellar astronomy. You could say the H-R diagram is to astronomers what the Periodic Table of Elements is to chemists.

Independently, Danish astronomer Ejnar Hertzsprung and the American astronomer, Henry Norris Russell, discovered that when they compared the luminosity with the type of light that was observed from stars, there were many patterns that emerged.

In 1905, Hertzsprung presented tables of luminosities and star colors, noting many correlations and trends. In 1913, Russell published similar data in a diagram. It is now called the Hertzsprung-Russell diagram in honor of these two pioneers. Russell noticed that almost 90 % of the stars fell along a diagonal ribbon that stretched from the top left to the bottom right of his diagram. The stars that fell onto this diagonal ribbon were classified as being on the "main sequence," of which our Sun is a member (Fig. 1.12).

They also noticed that other groups of stars become evident, such as blue supergiants to the upper right and white dwarfs to the lower left of the main sequence. O, B, and A stars are sometimes called "early type" while K and M stars are said to be "late type." This stems from an early twentieth century model of stellar evolution in which stars were powered by gravitational contraction via the Kelvin-Helmholtz mechanism. It was thought that stars were initially hot and bright (early type) and gradually evolved to become cool and dim (late type). The Kelvin-Helmholtz mechanism is an

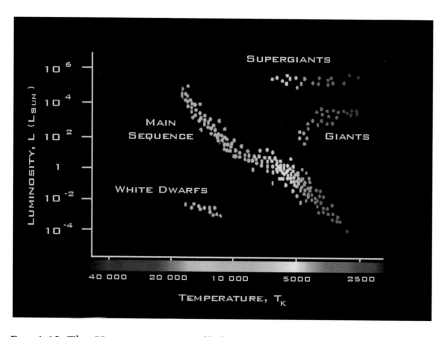

FIG. 1.12 The Hertzsprung-Russell diagram showing the main sequence (*diagonal line*) and some of the main groups of stars that lie off of it (Image credit: www.ie.ac.uk)

astronomical process that can be used to infer the lifetimes of the Sun and stars no older than a few tens of millions of years. A variety of subsequent scientific data points to a much older Earth and wider universe. Stars are now known to be powered by thermonuclear fusion. The lifetime of a star is strongly linked to its mass at birth. The more massive the star, the shorter its lifetime. The hottest stars burn themselves out in only a few million years, while the smallest can shine for hundreds of billions of years.

Intermediate-sized stars like the Sun end their days as bloated red giants, before shedding their outer layers to the great dark of interstellar space. What remains is a very dense white dwarf star, which cools off slowly over the eons. Stars larger than about 10 solar masses end their days much more violently, blowing off much of their matter in spectacular supernova explosions and leaving behind a tiny, spinning globe, no larger than a city, a super dense neutron star. After they go supernova, the cores of the very largest and rarest of stars collapse out of existence altogether to become black holes.

A General Note on Space Telescope Orbits

Launching a vehicle into space is an enormously costly venture, so deciding where to locate an observatory must inevitably be a compromise between astronomical advantages and the payload capabilities of the launch systems. Purely from the standpoint of the energy required for launch, the "low Earth orbit" (at an altitude of about 500 km) is by far the most economical. Anywhere lower than that would enable drag in the residual atmosphere to cause the spacecraft to fall back to Earth in only a few months, before burning up in the atmosphere.

In order to fully exploit the impulse supplied by Earth's rotation, the orbit should be placed nearly equatorial. This is the reason why launch sites are ideally situated at low latitudes—28.5 degrees for Cape Canaveral (USA) and just 5 degrees for European launches in the French Guyana. Both sites are located on the eastern side of a continent so that the launcher will fall into an ocean in the event of a launch failure. In contrast Baikonur, the Russian launch site, is situated inland at a latitude of 45 degrees because that country's coastal shores are not far enough south.

The period of a low Earth orbit is given by Kepler's third law, which states that the square of a satellite's period is proportional to the cube of its radius. At an altitude of 600 km, for example, a satellite completes one orbit in just 96 minutes!

There are other advantages of locating a spacecraft in a low Earth orbit, besides cost, that may not be immediately obvious. For one thing, Earth's magnetosphere in low Earth orbit protects the spacecraft from cosmic rays and the particle wind that emanates from the Sun; this greatly increases the longevity of the spacecraft and long-term astronomical observations they are carrying out.

That said, there are also several disadvantages to placing an observatory in low Earth orbit. For instance, roughly half of each orbit is actually unusable, owing to the fact that our planet occults the sky field under observation. What is more, telescopes need to be protected from the strong sunlight reflected from Earth during the orbital passage. For example, even with a 4-m long baffle in front of it, the Hubble Space Telescope cannot observe below 80 degrees from the horizon of the sunlit side of Earth. In addition to this, heat from nearby Earth creates a strong 'parasitic' background that is very detrimental to infrared observations, and this effect can only be eliminated by placing the spacecraft considerably further from Earth.

Thus, a much better solution is to locate orbiting astronomical observatories at the second Lagrangian point, L2, of the Sun-Earth system, where all these problems completely disappear. A Lagrangian point is a location in space where the combined gravitational forces of two large bodies, such as Earth and the sun or Earth and the moon, equal the centrifugal force felt by a much smaller third body. The interaction of the forces creates a point of equilibrium where a spacecraft may be "parked" to make observations. Travel time from Earth to L2 is about 100 days, and the launch can be assisted by passing close to the Moon. As we shall see, several observatories, either already in orbit or due for launch in the immediate future, are located at L2, especially those that image the universe at infrared wavelengths.

Another solution is to launch an observatory at escape velocity—that is, with just enough acceleration to escape the pull of Earth's gravitational field—then let it slowly drift away from Earth. This is a particularly economical way of launching, since no orbital insertion maneuver is required. The observatory will then assume a so called 'Earth trailing orbit'. The drawback with this approach is that the observatory would gradually drift away from Earth at a rate of about 15 million km per year. This is because of the uncertainty about the exact amount of impulse the launch system really provides. To avoid the risk of the spacecraft falling back to Earth, the launch impulse margin must be positive. The problem is that after several years, the observatory may float so far away that communications with it becomes problematical, if not impossible.

Now that we have discussed some of the underlying science behind the electromagnetic spectrum, the nature of atomic spectra, and the lives of stars, we can begin to fully explore humanity's progress in understanding the universe around us. It is an allegory filled with ingenious solutions to challenges and very surprising findings. We begin by exploring the most famous of all space observatories covering the visible and near ultraviolet regions of the EM spectrum—the Hubble Space Telescope.

2. The Hubble Space Telescope: Trials and Triumphs

Mention the words 'space telescope' to someone, and chances are the name 'Hubble' will trip off his or her lips. For nearly three decades, the Hubble Space Telescope (HST) has, more than any other telescope in history, revolutionized our understanding of the cosmos and our place within it. Yet, it was no plain sailing for the famous telescope, as we shall explore in this chapter.

Although HST was not launched until the 1990s, the impetus for its development actually began during the Cold War back in 1959; just one year after the Soviets launched their famous Sputnik probe. The then-chairman of the Space Science Board of the National Academy of Sciences, Lloyd Berkner, solicited practical suggestions for space-based projects to follow those undertaken in the International Geophysical Year that lasted from July 1, 1957 through December 31, 1958.

The response was encouraging, with more than two hundred proposals being received. These were subsequently studied by NASA and its advisory committees and used to refine a space-based observatory program. After such an extensive consultation, it was agreed that the launch of a large Orbital Astronomical Observatory was a priority in the following decades.

The dream of launching anything as sophisticated as HST had to begin with much more modest endeavors, to verify the 'proof of concept' as it were. In this capacity, astronomers first endorsed the launch of a string of smaller telescopes covering a variety of wavebands (including visual) before concentrating on a truly large space-based observatory. While these smaller missions were ongoing, Boeing received a contract from NASA to initiate the outline design of a 3-meter (120 inches) aperture instrument simply called the Large Space Telescope (LST). This size of telescope was deemed the maximum that could be successfully launched into space using NASA's most powerful launcher, the Saturn V launcher,

© Springer International Publishing Switzerland 2017
N. English, *Space Telescopes*, Astronomers' Universe,
DOI 10.1007/978-3-319-27814-8_2

which was first developed for the famous Apollo manned missions to the lunar surface. At this early stage, it was also anticipated that the telescope would be developed as part of a manned orbiting space station, a vision widely shared by space scientists in the post Apollo era.

A panel of 100 scientists and engineers brought together by the Space Science Board and NASA gathered at Woods Hole in Massachusetts in 1965 to discuss possible future space missions. At this meeting, a strong consensus of opinion was formed on the development of LST and a commitment was expressed to develop new technologies to make such a space telescope possible. The science priority of LST was envisioned to be in the discipline of cosmology, where it would be used to measure the distances to galaxies too far to be imaged from the ground and redefine the Hubble constant and the age of the universe (see Box 1 for an overview of this science).

Box 1: Hubble's Law and the Expansion of the Universe

In 1929, the American astronomer, Edwin Hubble, published the results of his studies on the red shift of a number of distant galaxies. His work was based at the then state-of-the-art 2.5-meter reflector atop Mount Wilson in California. Based on his observations of 24 very bright, distant stars (Cepheid variables), Hubble was able to state a new relationship;

$$v = H_o \times d$$

where v is the relative velocity of the galaxy, d is the distance of the galaxy in meters and H_o is Hubble's constant in units of per second (s^{-1}). During Hubble's day, the constant was calculated to be 1.62×10^{-17} s^{-1} but over the decades, after the accumulation of a much larger sample of galaxies, the Hubble constant was revised to its current value of 2.34×10^{-18} s^{-1} (Fig. 2.1).

A possible explanation for this relationship is that one event caused the creation and emission of all the matter in

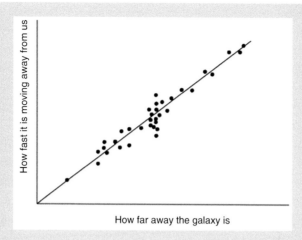

FIG. 2.1 Graph showing Hubble's Law (Image by the author)

the universe from a single point. This matter expanded exponentially, creating the universe as we know it today. Objects moving quickly are at extreme ends of the universe and relatively slower moving objects are closer to the 'center'. This would mean that observers like ourselves would see distant objects moving away from us. This recession velocity increases with distance, so that we would be moving away from slower objects nearer the center. From our observations, all galaxies appear to be moving away from us.

The Hubble constant has further use. The law suggests that the universe is expanding and if the rate of expansion is assumed to be fixed, then the Hubble constant may be used to determine the age of the cosmos.

Time = distance (d)/ speed (v), so Hubble's law gives $d/v = 1/H_o$.

Thus, $1/H_o$ yields the age of the universe. All recent measures place the age of the universe at around 13.8 billion years.

One of the biggest problems with getting LST up and going was to persuade a generally skeptical public that such a mission was worthwhile from an economic and scientific standpoint. In this capacity, the National Academy of Sciences Ad Hoc Committee on the LST convened a meeting in April 1966.

Chaired by the distinguished astronomer Lyman Spitzer, this committee cogitated on the relative advantages of sending a large astronomical telescope into space, as well as investing in a greater number and diversity of ground-based telescopes. By 1969, NASA officials had been convinced that there were sufficient funds to justify both approaches. In particular, the so-called Greenstein Report provided the cost estimate of launching a 3-meter class telescope into space—about $1 billion spread over a decade of time. Despite the Greenstein Committee's optimistic projections, Congress was less enamored by the project and refused to back the LST program. Furthermore, the approval of the Space Shuttle Program by NASA in 1972 meant that the most long-term commitments to space science had to be totally re-evaluated. This impelled astronomers Lyman Spitzer and John Bachall to take the opportunity and push the LST program forward since the Shuttle would provide new upgrades and replace components that a large orbiting astronomical observatory would eventually need.

A significant breakthrough was achieved in 1974, when Bachall and Spitzer persuaded a panel of 23 members of the Greenstein Committee to back the following statement:

> In our view, Large Space Telescope has the leading priority among future new space science missions.

As scientific advances progressed, any technological doubts were put to rest by new advances in both science and engineering. Thus, what started out as a 'pie in the sky' venture gradually transformed LST into a 'glowing endorsement' for both scientists and US Congress members. So, in August 1974, the LST program successfully surmounted the planning phase to begin phase B of its operations, but other problems still beset the project.

NASA needed additional funding from other nations to make the space mission a technological reality. In particular, the European Space Research Organisation (ESRO) was consulted to provide one of the instruments and other pieces of hardware that would fly with LST. The Marshall Space Flight Center was chosen as the main NASA site where LST would be built at an estimated cost of $517 million to $715 million.

The labor in the design of LST was divided up into three sections by NASA officials: the Optical Tube Assembly (OTA), the

Scientific Instruments (SIs) and the Support System Module (SSM). NASA found two potential contractors for the optical tube assembly. The next phase of making LST a reality began in 1977, when contracts were awarded to Lockheed Martin for the SSM, and Perkin-Elmer for the telescope optics. Around the same time, NASA officials chose the following instruments to accompany the Space telescope on its maiden flight:

The Wide Field Planetary Camera (WFPC), with James Westphal as principal investigator.
The Faint Object Spectrograph (FOC), with Richard Harms of the University of California as principal investigator.
The High Speed Photometer (HSP), with Robert Bless of the University of Wisconsin acting as principal investigator.
The High Resolution Spectrograph (HRS), with John Brandt of the Goddard Space Flight Center as principal investigator.
And finally, the Faint Object Camera (FOC) built and supplied by ESA.

It was widely anticipated that LST could be brought back to Earth for upgrading and refurbishment instead of being serviced in orbit. The Space Shuttle program opened up the new possibility of maintenance and even the complete replacement of scientific instruments. This meant that the number of ground-based tests the payload instruments had to be subjected to could be reduced in order to reduce costs. Within a few years however, it became ever more apparent that the design of the Space Telescope had to be improved in order to minimize the number of repair trips needed by the Space Shuttle. This called for a revision of the program's cost and by 1980, the estimates for the entire program were between $700 and $800 million, with the anticipated launch date postponed until 1985. In private, many of the senior project managers assigned to the LST program suspected that the mission would be decommissioned by members of Congress.

Towards the end of 1982 more trouble met the LST program. PerkinElmer announced that their part in the LST was seriously underfunded and requested an additional $450 million to complete the telescope within the original schedule. Furthermore, the ambitious mission to launch LST suffered from poor management. These events impelled Samuel Keller,

NASA's deputy associate administrator, to re-evaluate the entire LST program, concluding that the project was at least six months behind schedule and would need an additional $100 million to bring the LST to fruition.

What followed was a new and invigorated discussion with Congress that resulted in a larger budget of $1175 million, a 70% increase in cost over the original budget plans! The launch date was changed to the end of 1986 by NASA. Moreover, the LST was renamed the Hubble Space Telescope (HST), after the American astronomer, Edwin Hubble (1889–1953), who discovered the expansion of the universe (see Box 1). Finally, while the original LST was designed to have an aperture of 3 meters, HST had to be reduced to just 2.4 meters (Fig. 2.2).

It was widely anticipated that the space telescope would be ready in time to assist Voyager 2's nearest approach to Uranus in 1986, as well as to observe Halley's Comet, during its perihelion passage a month later. But it was not to be. A considerable number of problems attended the American part of the program, so much so that it became incumbent upon senior NASA officials

FIG. 2.2 HST's mirror blank being ground (Image credit: NASA)

to take a more active role in overseeing the HST project. Accordingly, a new project manager, John Welch, was appointed and NASA contracted BDM Corporation to undertake systems engineering of various components of HST's design. With a considerably enlarged budget, NASA made it a priority to greatly reduce the risk of the entire program by increasing both the number of spare components as well as the number of replacement units that could be placed in orbit. During this restructuring period, NASA finalized its commitment to undergo all repairs of HST from Earth orbit rather than bringing the telescope back down to Earth.

Though all parties redoubled their efforts to bring the project to fruition in the years following the 1983 hiatus, disaster struck. On January 28, 1986, the Space Shuttle Challenger exploded, putting the launch of HST on ice. In addition to this, a string of new problems were uncovered during the equipment testing under vacuum conditions. The cold (pun intended) reality was that HST would have missed its launch window anyway. Thus, a new launch date was set for November 1988, but even then, further delays in restoring the Shuttle Program meant that HST could not have been launched into space before 1990 (Fig. 2.3).

FIG. 2.3 Edwin P. Hubble (1889–1953), the famous American astronomer after which the Hubble Space telescope was named (Image credit: recherche-technologie.wallonie.be)

The Innards of HST

On April 24, 1990, after all setbacks had been resolved, the Hubble Space Telescope blasted off from Cape Canaveral on board the Space Shuttle. At launch, its total cost to NASA had skyrocketed to about $2000 million and $350 million to ESA. At the heart of the telescope was an f/24 Ritchey-Chretien reflector, with a primary mirror of 2.4-meter aperture (a little less than the 3-meter instrument originally planned) and weighing in at nearly one metric ton. In launch mode, the entire HST—sans its aperture door, solar panels, and antennae—had a length of 13.1 meters and was 4.4 meters in diameter, tipping the scales at 11.6 tons. The telescope was powered by solar panels with a power consumption of 2.8 kilowatts. It was placed in a 610 kilometer altitude orbit inclined at 28.8 degrees and had an anticipated minimum lifespan of 15 years. NASA's long-term plan was to include a refurbishing/repair mission every three years by astronauts using the space shuttle.

Up until 1990, the largest ground based telescopes could achieve a resolution of 0.5 arc seconds (without adaptive optics) in the best seeing conditions, and about 1.0 arc seconds for about 2000 hours per year. The HST was expected to have a resolving power of about 0.1 arc seconds in the visible region of the EM spectrum and an impressive 0.015 arc seconds at ultraviolet wavelengths (121.6 nm). The amount of time that HST could engage in observations of the universe came to about 7000 hours per year. The telescope was expected to operate at UV, visible and near infrared wavelengths, covering wavebands from about 110 nm right up to 1.0 millimeter, giving it a far wider waveband range than any Earth-based telescope. This spectral range was not possible when HST was initially launched since the instrumentation capable of imaging at infrared wavelengths was not available when it first entered the vacuum of space.

Great advances in positioning technology meant that HST could undergo long-exposure imaging with a pointing accuracy of 0.01 arc seconds and remain within 0.007 arc seconds of that position for a full day! With such high resolution and exceptional pointing accuracy, it was expected that the telescope would be able to detect stars as faint as visual magnitude 27 after an exposure of just four hours. However, by extending the exposure time to 24 hours, objects as faint as magnitude 29 could be detected!

The Wide Field and Planetary Camera (WF/PC) placed on board HST was developed by engineers at JPL and consisted of eight CCDs arranged in two groups of four. The first CCD group was devoted to imaging in wide field mode (f/12.9) and the other arrays were used to image high resolution targets at f/30. The wide field CCD array had a 2.7 × 2.7 arc minute square field, made up of 1600 × 1600 pixels each covering an area of just 0.1 arc seconds. In contrast, the second planetary CCD of the WF/PC array was dedicated to imaging a much tinier field some 69 square seconds, with each of its 1600 × 1600 pixels covering just 0.043 arc seconds. The entire CCD array was cooled to a chilly −100 °C to optimize its signal-to-noise ratio. The individual pixels had to be coated with a special phosphor material called coronene to improve their efficiency in the UV region of the spectrum, as well as to enable a spectral range from 115 nm (UV) right up to 1.1 microns (infrared).

The Faint Object Camera (FOC), which was built by German firm Dornier Systems and commissioned by ESA, was fitted with an f/48 and f/96 camera system. The former had a field of view of 22 square arc seconds and came equipped with a variety of filters, prisms and diffraction gratings for high resolution spectroscopy. The f/96 camera was likewise designed to use polarizing filters and was also fitted with a special coronagraph to allow very faint objects to be imaged just one arc second away from another object up to 17 magnitudes brighter.

In the very highest resolution mode, which was necessarily confined to the ultraviolet region of the EM spectrum, the f/96 camera was transformed into an f/288 system that covered a field of view of just 4 arc seconds and a pixel size of 0.007 arc seconds. A team at British Aerospace designed the detectors for the FOC, which consisted of very high efficiency photomultipliers. These in turn were coupled to television cameras that, remarkably, could detect a single photon of light! In addition, the full 1024 × 1024 pixel could be processed in either 8-bit or 16-bit mode.

Two other instruments coupled to HST included the Faint Object Spectrograph (FOS) and High Resolution Spectrograph (HRS) both of which were fitted with an array of diodes to measure radiation intensities at various wavelengths of the spectrum. HRS was designed and constructed by Ball Aerospace and could produce both medium and high resolution ultraviolet spectra of stars as faint as magnitude 19. Martin Marietta designed the FOS, which was able to capture spectra of significantly fainter stars with magnitude as low as 22.

The High Speed Photometer (HSP), the simplest and cheapest of HST's on-board instruments, was designed by a team of engineers based at the University of Wisconsin and, remarkably, contained no moving parts. This extraordinary instrument was capable of detecting minute changes in brightness in objects as faint as magnitude +24 and with a time resolution of just 0.0.016 milliseconds.

Over the first few days, after arriving in its assigned orbit, HST opened up its solar panels and deployed its high gain antenna. The cargo bay of the Shuttle was then slowly opened and the famous telescope carefully coaxed into position. Yet even at this stage, a few technical hitches delayed the telescope from seeing first light until May 20. With a sense of great anticipation, HST's first target was the open cluster NGC 3532, which also contained a double star with angular separation of just one arc second. The double was clearly resolved although, puzzlingly, the two stars were not clearly defined points of light. And while some NASA officials were visibly relieved, others immediately suspected that there was something awry with the images.

On the next evening, Roger Lynds of the NOAO and a member of the WF/PC team boldly suggested that the HST had a pretty bad dose of spherical aberration. Chris Burrows, an optics expert with ESA, agreed with Lynds, adding that as well as spherical aberration, the telescope images were showing significant levels of coma. After attempting a few clever maneuvers, like moving the primary mirror towards the secondary or using the actuators to change the shape of the mirror ever so slightly, the images could still not be improved. When the first images from the FOC were processed, images showed exactly the same effect. Over the next couple of days, engineers realized that the spherical aberration showing up in HST's image could not be fixed using hardware on board HST (Fig. 2.4).

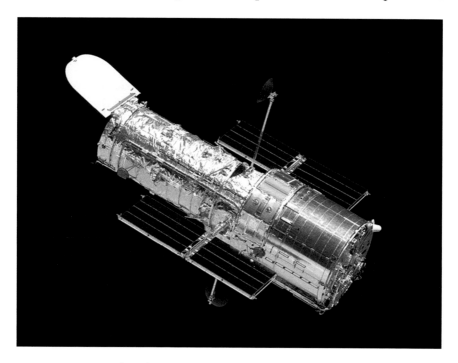

FIG. 2.4 HST in orbit above the Earth (Image credit: NASA)

On June 27, 1990, Ed Weiler, the chief scientist with the HST program, was given the unenviable task of announcing to a room full of curious journalists that the newly launched $1.5 billion observatory's supposedly flawless 2.4 meter primary mirror was misshapen because it was ground to the wrong shape, and so was unable to bring starlight to a crisp focus. In theory, the mirror ought to have brought 70 % of the light incident upon it to the same focal point, but HST could only manage a paltry 10 to 15 %.

Moving from the center to its edges, the mirror was discovered to be too shallow by about 2 microns. In and of itself, this was only a tiny fraction of the width of a human hair, but it was enough to destroy the definition of the famous space telescope. From an optical standpoint, the parabolic shape of the mirror was wrong! This tiny imperfection made NASA's greatest telescope the butt of jokes all over the world and the subject of great consternation on Capitol Hill.

After a long and protracted investigation by officials, the root cause of the calamity was elucidated. A technician had inadvertently inserted a small 3 mm diameter washer into a device called a null corrector, an instrument employed to check the mirror's shape during its production a few years earlier. In retrospect, there were tell-tale signs that something was wrong during the mirror testing phase of project, but these were completely ignored. What is more, performing a thorough optical test on the primary mirror was not undertaken because of efforts to minimize costs. In addition, it was decided that performing these tests would compromise the cleanliness of the telescope. Skipping added testing would thus reduce the risk of biological contamination.

Understandably, Weiler was totally devastated by the findings. "If you had polled all the engineers and scientists at Cape Canaveral the night before the launch for the top ten concerns they had," he said, "what could break on Hubble or what wouldn't work on Hubble, I would bet my house and a lot more that not one of them would put on their list the mirror's wrong shape and we've got spherical aberration. Nobody worried about that because we were assured by the optics guys that we had the most perfect mirror ever ground by humans on Earth."

Remarkably though, the misshapen HST primary mirror was still used to undergo some cutting edge science. One of the most ambitious goals of the astronomical community was that HST would be capable of imaging planets orbiting their parent stars via the Faint Object Camera (FOC). Unfortunately, the defect to Hubble's mirror precluded the use of its coronagraph, so these exoplanets couldn't be detected. Significantly though, the hobbled HST could still be used to image the gas and dust clouds surrounding young stars such as Vega and Beta Pictoris—with the aim of following up earlier work conducted by the Infrared Astronomical Satellite (IRAS). In 1991, a team of astronomers led by Albert Boggess of the Goddard Space Science Center employed the High Resolution Spectrograph to examine the spectrum of Beta Pictoris at UV wavelengths. These studies demonstrated that the gas present in the dust cloud was moving inwards towards the star at speeds on the order of 50 kilometers per second. The same team also uncovered clumps of gas that many now think are new planets in the process of forming (Fig. 2.5).

Throughout the summer months of 1990, NASA engineers precisely mapped the shape of HST's mirror, discovering it had

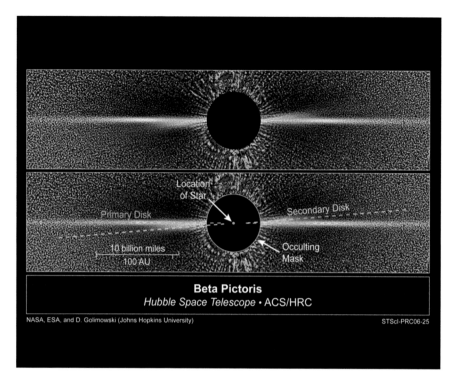

Fɪɢ. 2.5 HST images of the dusty disk surrounding the star Beta Pictoris (Image credit: NASA)

been ground just as perfectly as advertised but to precisely the wrong shape! While this investigative work was going on, NASA engineers picked their brains in order to arrive at possible fixes. These included everything from sending an astronaut crew to HST to replace the secondary mirror with another one, installing a circular diaphragm around the periphery of the tube, thereby reducing the aperture and improving the focus by blocking out the outer regions of the primary mirror which had the greatest defect. Before all that happened however, NASA had to 'officially' offer an admission that the telescope mirror was misshapen, and it was incumbent upon Weiler to make that announcement.

When the whole world was aware of HST's problem, it was the NASA optics experts' responsibility to arrive at a workable solution to the problem. It was John Trauger, the principal investigator with the WF/PC, who first suggested a way around the problem. Analyzing the images from HST, optical engineers employed a computer assisted method known as deconvolution to design 'corrective glasses' with a prescription that introduced aberrations that were exactly the opposite in sign from those ground into the mirror. By a stroke of good fortune, an exact replica of the WF/PC that was built several years earlier could be called into service to provide that correction. But that would only work if a set of small and carefully designed mirrors were positioned in front of the original Hubble instrument, the Corrective Optics Space Telescope Axial Replacement (COSTAR).

In August 1990, the FOC was instructed to carry out a 28 minute exposure of SN 1987A, the supernova remnant located in the Large Magellanic Cloud, at a specific wavelength corresponding to the line of doubly ionized oxygen (OIII) at 500.7 nm. Such an exposure showed that the star was surrounded by an enormous egg-shaped structure measured to be about 1.6 arc seconds across. Initially, it was thought that this structure was just a ring of material that must have been ejected by the exploding star at some time in the past, and inclined at 43 degrees to our line of sight. The ring was set alight after it was irradiated with UV light some 240 days after the explosion. This meant that sometime between 1999 and 2002, the material from the supernova explosion was predicted to collide with the ring debris. What's more, these observations enabled astronomers to increase the accuracy of the distance to the LMC (which is now measured at 169,000 light years with an error margin of just 8000 light years).

The enormous light grasp and resolving power of HST allowed astronomer Francesco Paresce, based at the Space Telescope Science Institute (STScl), to peer deep into the center of the bright globular cluster 47 Tucanae. By nulling the light from the predominant red giant stars, the team found twenty one so-called 'blue stragglers' in regions close to the core of the cluster, which suggested that some stars must have come so close to each other within the dense cluster that they coalesced with each other in some sort of violent merger. Before these observations were made, it had been widely and commonly accepted that all globular clusters were of roughly the same age. But if blue stragglers had been produced either by stellar mergers or by some kind of mass transfer process, some of the globular clusters were being rejuvenated, even if they were strictly the same age as their red giant companions. Thus, the color of a star is not an absolute indicator of the age of stars in a cluster.

In 1991, Lowell Observatory astronomer, Jon Holtzman, and colleagues, utilized the WF/PC 1 to image the central region of the elliptical galaxy NGC 1275 at the heart of the Perseus galaxy cluster, finding what appeared to be about 50 blue colored globular clusters that were estimated to be only about 300 million years old. This was a very surprising discovery, which could only be explained if several fairly recent galaxy mergers took place to create NGC 1275. Globular clusters are not necessarily the 'old men' of the galaxy and could have formed much later in the evolution of the Milky Way Galaxy and other galaxies. In other galactic work, astronomers turned HST toward the active radio galaxy M87, which was thought to harbor a supermassive black hole at its epicenter.

As we have seen, more than 1000 astronomers used HST to gather images even in its flawed state but had to rely heavily on computer processing techniques to glean the best from the beleaguered orbiting observatory. But this was no panacea compared to going up and fixing the optics.

When John Trauger (JPL, Pasadena, California) learned that the optical problem had been traced to a misshapen mirror, he immediately saw a partial solution. Since 1985 he and his colleagues had been building a backup for Hubble's main scientific instruments, the Wide Field and Planetary Camera (WFPC) and it could easily be modified to negate the effect of the telescope's aberration.

Like the first one, the second camera would contain eight small Cassegrain reflectors. These relayed the f/24 beam from the telescope to a sensitive CCD camera, which changed the image scale to f/12.9 for the four wide field chips and to f/28.3 for the purpose of high-resolution planetary imaging. Trauger realized that the Cassegrains' dime-sized secondary mirrors could be given an amount of spherical aberration equal in magnitude but opposite in sign to the Hubble's, thereby bouncing corrected images to the CCDs.

Before new relay secondaries could be manufactured, scientists had to determine the right prescription. They did this in two ways; by studying actual images from both cameras and by examining the fault test device that caused the spherical aberration in the first place. The results agreed by more than 5 %.

HST's primary mirror was about 2 microns too flat at the edge, so the corrective optics had to be higher at the edge by the same amount. Since these mirrors were only a centimeter in diameter, it followed that they had to be very steeply curved. That said, achieving this figure was fairly easy to achieve. What Trauger and his colleagues could not foresee, and what turned out to be a much more challenging problem, was the task of aligning the optical components in such a way as to provide diffraction limited performance even in the presence of spherical aberration.

If the telescope and camera were incorrectly collimated by as little as 2 % of the relay mirror's diameter, then coma would have become the predominant optical aberration and HST would have produced even worse images than before. The existing latches and rails that positioned the camera in the telescope were simply not rigid enough to provide the level of alignment tolerance needed. So, Trauger and his associates were left with no better option than to adjust the camera's new optics in orbit. Such a project added to the expense of the mission so something else had to go.

After considerable reflection and deliberation, the team decided to remove four of the eight CCD channels. This meant that it could no longer alternate between two-by-two mosaics of f/12.9 (wide field) and f/28.3 (planetary) detectors. In contrast, the new camera could produce a single mosaic image comprised of an L-shaped trio of wide-field images as well as a smaller, high-resolution image sampled in the corner of the CCD array.

The pickoff mirror that directed the light from the telescope into the camera was now fully steerable, as were the fold mirrors that lined up the beams with the relay Cassegrains. The mirrors were positioned on three ceramic actuators, the heights of which could be lengthened or shortened by a few microns when a small voltage was set up across them. In this way, adjustments could be made to ensure that the alignment was spot on.

Despite the loss of half its CCD detectors, the new camera was significantly more sensitive than the old one, allowing shorter exposures to be made. WFPC-2 was an improvement on its predecessor in other significant ways too, like its ability to offer better performance at ultraviolet wavelengths. Trauger was satisfied that these innovations would bring HST's capabilities in line with what astronomers would be happy with.

Astronomers and NASA were understandably excited by the easy fix available through WFPC-2, but many still expressed concerned about the fate of Hubble's other instruments, none of which had planned replacements. To address this issue, Riccardo Giacconi, the director of STSc1 at the time, appointed a "strategy panel" of astronomers, optical engineers, and astronauts in August 1990 to explore strategies that could be adopted to fix the Faint Object Camera (FOC), the Goddard High Resolution Spectrograph, the Faint Object Spectrograph and the High Speed Photometer on board the orbiting observatory.

The panel aimed to establish a practical method that would fit into a program dedicated to the telescope's maintenance and servicing schedule. Murk Bottema of Ball Aerospace first suggested using mirrors similar to the WFPC-2 relay secondaries. Because they absorb UV radiation, ordinary glass lenses simply wouldn't have worked. To counter this, Bottema's idea involved the use of two mirrors working in concert to correct the aberration beam entering each of the scientific instruments. The first mirror would intercept light from a convenient spot in the telescope's field and direct it to a second mirror, which would cancel out the aberration before directing a corrected beam into the instrument.

Around the same time, Crocker discovered a way to direct the mirror images to the right places in the focal plane. During the development phase of HST, a 'dummy' scientific instrument was set up for testing. Consisting of little more than an empty box, the Space Telescope Axial replacement (STAR) could be substituted for any of the phone booth-sized instruments that were mounted parallel to the telescope's optical axis (that is, all instruments but the WFPC). Crocker suggested the removal of one of these instruments and the insertion of a modified STAR to house Bottema's Corrective Optics STAR (COSTAR).

The same panel of experts recommended that COSTAR replace the HIGH SPEED Photometer (HSP), which was Hubble's least used instrument. HSP team leader Robert Bless (University of Wisconsin) initially planned to include a miniature HSP made from spare parts with COSTAR. In the end, engineers assigned to COSTAR could not find a suitable light path to enable such a device to operate. Ball Aerospace completed COSTAR in two years and four months. The 290-kilogram marvel of technology contained five mirror pairs—two each for the FOS, FOC and one for the Goddard spectrograph, whose two apertures were close together. These mirrors had apertures in the range of 12 to 25 mm, some of which had steeply curved spherical surfaces to provide the necessary optical correction. As with WFPC-2, optical alignment had to be very precise, so provision had to be made to enable flight controllers to adjust each Mirror 1 in such a way as to aim it squarely at Mirror 2. All ten mirrors were installed on COSTAR and mounted on a sophisticated optical bench. A veritable marvel of miniaturization, the device was actuated using 12 tiny DC motors with ten mirrors, four arms, and innumerable sensors and connecting wires compacted into a space the size of a shoebox.

The installation of COSTAR ensured that Hubble's image quality would be good enough to concentrate 60 % of a star's light inside a radius 0.1 arc seconds, making it almost as good as the original 70 % aimed for by NASA. COSTAR wasn't perfect however. It increased the focal ratio (or magnification) of each of the corrected beams, and in so doing reduced the apparent field of view of the image. For example, the original f/48 and f/96 operational modes of the FOC now became f/75 and f/151, respectively. What is more, the two additional reflections reduced the amount of light reaching the instruments. In particular, the combined reflectivity of Mirrors 1 and 2 meant that only about 80 % of visible light and 60 % for ultraviolet light could be captured.

When HST's optical glitch was first uncovered, the pertinent question on everyone's mind was; why wasn't it remedied before it was launched? NASA could not afford to make the same mistake twice and so had WFPC-2 and COSTAR thoroughly checked over and over again, in different ways, in order to ensure that would both fit in the telescope as planned and perform as intended.

Remarkable though it may seem, the optical aberrations of HST were not the first problem to beset the famous orbiting observatory. From the outset of its days in orbit, the great telescope had vibrated violently as it traversed the day-night terminator. After careful study, engineers determined that the origin of these vibrations arose as a result of the spacecraft's twin solar cell arrays. Apparently, the giant panels (each 12 meters long and 2.8 meters wide) began to flap in response to the rapid heating or cooling that attended each terminator crossing, resulting in a noticeable jitter in the telescope's pointing that occurred for 5 or 10 minutes. Usually when the solar cells were quiet, the tracking sensors locked onto suitable guide stars with a precision better than 0.005 arc seconds which actually exceeded NASA's anticipated pointing accuracy (0.007 arc seconds). But once the jittering started, the telescope would be knocked off target, disrupting the intended observation and the recording of precious data. Indeed, throughout HST's first year in orbit, all observations that occurred during terminator passages had to be cancelled and re-scheduled. Eventually though, NASA flight controllers gained sufficient understanding of these vibrations to reprogram the HST's onboard computer and eliminate the disruptions by making slight adjustments to the rotation speeds of onboard flywheels.

After these bugs were ironed out, the telescope hardly ever lost its ability to lock onto targets. However, the much-increased complexity of the jitter reduction software cut into the already limited amount of computer memory available for data retrieval and analysis. More alarming still was the prospect that the persistent flapping of the solar panels would eventually cause mechanical stress (such as metal fatigue) in the booms that held the arrays together on the telescope. For example, if one of them were to pry itself from the body of the telescope, it would quickly drain the electrical power and cause the whole spacecraft to shut down. As luck would have it, the company that built the solar arrays, British Aerospace, had already been working on a spare set of panels as soon as this problem came to light. These new panels were planned to be installed in the mid to late 1990s, by which time the power output on the original panels would have degraded considerably owing to overexposure to the deadly environment of outer space. This time however, the new arrays were to be given additional thermal protection, and their construction was sped up so that they would be completed in time for the greatly anticipated 1993 shuttle repair mission.

The solar cells were arranged in blanket-like layers that could be extruded in opposite directions from a drum, and made to move along by thin rods called BiSTEMs. Constructed from two flexible, semiconductor ribbons, they slid together as they were extended from their storage cassettes, similar to an ordinary household tape measure. The blanket ends were affixed to a bar placed between the two BiSTEMs. The project manager at British Aerospace, Michael Newns, figured out that these BiSTEMs, as well as the drum and the attachment between the blankets and bar, had all contributed to the heat-induced jitter, since each would have the potential to stick or suddenly become free to slip as the temperature alternated between hot and cold. The improvements that were built into these new arrays consisted of bellows fashioned from aluminized Teflon that were used to insulate the BiSTEMs, a braking system for the drum, and an attachment mechanism for the blankets.

While the Shuttle mission prioritized correcting the telescope optics and the jitter of the solar array, a potentially more serious threat came to the fore. Three of HST's six gyroscopes failed, leaving a key component of the pointing-control system without any replacements. These gyroscopes were designed to sense the telescope's turning rate, three of which were necessary to calibrate the motions around HST's three axes. The gyroscopes were packaged in pairs inside three rate-sensing units (RSUs), each an associated electronics control unit (ECU) to regulate them.

In a lucky twist of fate, one gyroscope had failed in each of the three RSUs so the telescope continued to operate. Two of these shut downs had already been linked to faulty RSU circuitry. However, two of the gyroscopes that remained operational utilized the same circuitry. The third and final failure apparently occurred when a fuse blew elsewhere in the gyro power system, shutting down the ECUs as a result. So the replacement of two RSUs and two ECUs, together with an upgrade to more sophisticated circuit breakers, would leave HST with six operational gyroscopes.

Even if HST had perfect optics and rock stable solar panels, NASA would still have to dispatch a repair mission to replace any malfunctioning gyros to allow the telescope to continue making observations of scientific value. It might have been possible to control HST using a pair of gyroscopes, magnetometers and sun sensors, but this couldn't happen without upgrading the software controlling the telescope's moment-to-moment operation. Any such upgrading was beyond the remit of the repair time leading up to the planned launch date. Addressing a team of astronomers, servicing program manager, Kenneth W. Ledbetter, after calling attention to this precarious situation quipped, "If you know any gyro heath chants, now would be a good time to say them!"

HST was not without its fair share of electronic problems. Two out of the six memory modules installed in the spacecraft's main computer had malfunctioned, as did the prime unit which oversaw the exact positioning of the solar cell arrays. What is more, the Goddard spectrograph with its low voltage power supply, an image intensifier in the Faint Object Camera and both magnetometers started to behave erratically and even failed to work on some occasions. After some investigation, the cause was found to be broken solder connections and/or failed components.

To repair the main computer, the astronauts were forced to include a co-processor module with a great deal of additional memory. If that failed, however, repairs would take considerably longer and would overstretch an already cramped space-walk schedule. In addition to the co-processor module, HST would get new and improved solar-array drive electronics, as well as a relay box to circumvent the spectrograph's balky power supply. New magnetometers would also be installed during the repair mission.

Three space walks were planned for the Hubble repair mission scheduled for January 1993. However, the number of problems continued to mount for the famous space telescope, resulting in further launch delays due to needed repairs. By March of 1993, the number of space walks increased from three to four, then to five in June. At that point, the Space Shuttle Endeavour, which became the newest member of NASA's fleet, was the only orbiter capable of making the flight, as it alone could carry the prerequisite amount of fuel, air and water to successfully carry off the mission.

All the sophisticated cargo required for the repair mission had been delivered to NASA by late August 1993, and some had even been dispatched to Cape Canaveral for incorporation into the payload bay of Endeavour. NASA had divided the cargo into two main classes. The primary payload consisted of COSTAR, WFPC 2, new solar panels, gyro pairs with fuses and their attendant electronics, a magnetometer and the electronics that ran the solar panels.

The Secondary items were designed to improve HST's scientific capabilities, and included the Goddard spectrograph repair kit, the computer co-processor to reduce the changes of failure, another magnetometer and a second set of gyro electronics. Planning the sequence of events that would successfully restore HST to good working order was no easy task and, inevitably, had to include compromises between payload priorities and the time available for each spacewalk. It was all a massive exercise in risk assessment. Reviewing the scientific literature of the day, the pressure on NASA to deliver in the aftermath of the fiasco was palpable to say the least!

The all-important COSTAR apparatus, designed and built by Ball Aerospace over the course of three years, together with WF/PC 2, were both ready to be included on the first anticipated HST servicing mission, which was pushed forward to December 1993. In the hours leading up to the spacewalk to rejuvenate a myopic Hubble, a 50-strong army of astronomers and engineers nervously congregated around a large television screen as that first picture was downloaded for analysis. "That was the first moment I knew we had fixed it," Weiler recalled. "The first picture had a star right in the center. It was only that star, but it was crystal sharp clear and the crowd went crazy! I thought the then NASA Administrator, Dan Goldin, was going to fire me the next day but instead, he congratulated me."

NASA was understandably reserved about releasing the first light results of the most famous servicing mission in history, deciding instead to collate more data that would silence a skeptical press and public, demonstrating once and for all that the most expensive telescope in the history of the world was worth every dime lavished upon it. So, slowly but surely, over the Christmas period of 1993 and extending into the first few days of January 1994, about a dozen new images from HST were carefully analyzed by mission scientists (Fig. 2.6).

NASA finally arranged a press conference at Goddard on January 13, 1994, which coincided with an American Astronomical Society meeting in nearby Washington. The conference even attracted the attention of Barbara Mikulski, Maryland senator, who was shown the 'before' and 'after' images at first hand. "My god," she exclaimed, "it's like putting my glasses on!"

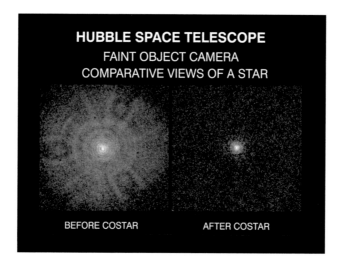

Fɪɢ. **2.6** Showing a star image obtained before and after the servicing mission (Image credit: NASA)

So, in the end, the $50 million COSTAR as well as the $23.9 million WF/PC 2 upgrade saved HST from becoming the great big techno turkey in the sky the press wanted it to be. But soon, HST's intrepid mission was ready to push the envelope out on our knowledge of the universe, as well as our place within it. This will be the subject of the next chapter.

3. Hubble: The People's Telescope

The design and launch of the Hubble Space Telescope (HST) was a victory for human ingenuity. Never since the time of the American Apollo program has the genius and diligence of countless individuals across a dozen disciplines dovetail so well together in the creation of the world's premier telescope. Over the years, whole teams of scientists, engineers, technicians, craftsmen and astronauts devised ways to transform a flawed, decrepit observatory into a state-of-the-art research facility. Subsequent servicing missions to the space telescope confirmed this in the years after its launch.

From its inception, it was clear that the astronomy community did not want NASA to coordinate the day-to-day running of the telescope. As a result, various consortia were set up to manage the hour-to-hour operation of the various instruments on board HST. The most influential among these groups was the Association of Universities for Research in Astronomy (AURA), the body which was eventually selected—in 1981—to run a brand-new institute dedicated to overseeing HST. This set the stage for the development of the Space Science Institute, situated at Johns Hopkins University in Baltimore, Maryland, USA.

HST enjoyed a total of five servicing missions in the years 1993 through 2009, each progressively increasing its power and versatility. In time, it would earn its stripes as the People's Telescope. The first servicing mission after the 1993 fitting of WF/PC 2 and COSTAR arrived in February 1997. Space shuttle Discovery carried a team of seven astronauts to replace the two aging spectroscopy instruments—STIS and NICMOS—with more up-to-date instrumentation. These new spectroscopy packages broadened Hubble's spectral sensitivity range, allowing it to see hitherto hidden parts of the EM spectrum. It was no mean feat, though, requiring two full days work with 150 tools and a total of five space walks. The same crew installed an updated gyroscopic sensor to improve the telescope's pointing accuracy, swapped out a data recorder and fit thermal insulation panels.

© Springer International Publishing Switzerland 2017
N. English, *Space Telescopes*, Astronomers' Universe,
DOI 10.1007/978-3-319-27814-8_3

A third servicing mission was conducted by astronauts on board the Space Shuttle Discovery in 1999 to replace one of the guiding sensors and all the gyroscopes needed to point the telescope. A new solid-state digital recorder was also fitted. Three years later, the Space Shuttle Columbia delivered the Advanced Camera for Surveys (ACS) which replaced the FOC. In addition, new solar panels were fitted and NICMOS received a new cooling system. Finally, in 2009, Space Shuttle Atlantis replaced both COSTAR and WF/PC 2 with the Cosmic Origins Spectrograph (COS) and Wide Field Camera (WFC) 3, respectively. In addition, ACS and STIS was repaired, more new gyroscopes fitted and HST's batteries replaced with fresh ones. A capture mechanism was also added so that in the future, a tug could latch onto HST and de-orbit the telescope safely at the end of its life.

There is virtually no aspect of astronomy and astrophysics where HST has not contributed something new and significant. Its fame rests firmly on the steady stream of discoveries and observations in everything from planets, moons, asteroids and comets in our solar system to the furthest reaches of the cosmos, capturing the feeble light from galaxies emitted within a half billion years or so of the Big Bang, the birth of the universe. Indeed, together with Galileo's spyglass and the 100-inch Hooker reflector atop Mount Wilson, HST goes down in history as one of the most productive astronomical telescopes of all time (Fig. 3.1).

As soon as the first corrective optics package was installed on HST back in 1993, the telescope was ready to explore one the biggest questions posed by modern cosmologists: how old is the universe? To answer this question, astronomers had to refine the distances to galaxies and how fast they are flying apart. Then, by running the movie backward in time, they established the likely age of the universe. Edwin Hubble used the 100-inch Hooker reflector to show that the universe was expanding, with the more distant galaxies moving away faster than those closer to us. The venerable astronomer used old, pulsating stars known as Cepheid variables to estimate the distance to nearby galaxies, but HST was able to greatly refine those distances and extend its reach into galaxies much further afield than any survey before it. As a result, we now have the best figure yet for the age of the universe—13.8 billion years old with a precision of 3 %. Future

Fɪɢ. **3.1** HST images of 'toppled over' Uranus with its extraordinary ring system. WF/PC 2 revealed features deep inside its atmosphere never seen before (Image credit: NASA)

observations by HST will enable cosmologists to reduce this uncertainty to just 1 %.

HST astronomer Adam Riess used the famous space telescope to hunt down extremely distant Type Ia supernovae—new standard candles—to study the expansion rate of the universe over cosmic time. Standard candles are astronomical objects, the properties of which are very well known, and can be used to determine distances to objects located at cosmological distances. This was a very hot topic back in the late 1990s as two teams of researchers— the Supernova Cosmology Project and the High-Z Supernova Search—engaged in an intense but friendly competition to measure how the cosmic expansion might have changed over the age of the universe. The widely anticipated result predicted that the expansion was slowing down, which is what one might expect in the absence of any force other than gravity. The details—how fast or how slow it is decelerating—are critical for understanding the evolution of the cosmos and predicting its eventual fate.

To grapple with this, astronomers needed to study galaxies at much greater distances and far deeper in the cosmic time. Cepheid variable stars could not be used because HST could not distinguish individual stars at such enormous distances, but Type Ia supernovae could be seen across the visible universe. Type Ia supernovae occur when a white dwarf in a binary star system accretes enough mass from its companion star to exceed 1.4 times the mass of the Sun, a tipping point known as the *Chandrasekhar Limit*. When that happens, a sequence of nuclear reactions rip through the star, triggering a titanic explosion that is five billion times brighter than the Sun! Astronomers believe Type Ia supernovae all happen the same way, producing brilliant flares with similar, predictable light curves. That enables researchers to calibrate the light of a Type Ia supernova several billion light years away to determine how far away it must be.

If the expansion of the universe is slowing down, one would expect the light from Type Ia supernovae to be brighter and thus closer than they would be in a universe expanding at a constant rate. Much to the surprise of the scientific community however, the astronomers discovered that Type Ia supernova in distant galaxies are much fainter than expected, implying that they are farther away than they would be in an decelerating universe. The unavoidable conclusion was that the expansion of the universe is speeding up, not slowing down. Researchers eventually concluded that some sort of repulsive force—'dark energy'—must be powering this universal acceleration, a discovery that ranked as one of astronomy's greatest achievements.

HST provided a crucial observation in this discovery. Riess, who was a member of the High-Z Supernova Search, used the telescope to find some of the most distant Type Ia supernovae ever discovered. Analyzing the ancient starlight, Riess found that that they were brighter than they should have been if the acceleration seen in younger targets are constant. The data indicated that the universe did, in fact, decelerate over the first eight billion years or so of its life. Then it started speeding up.

Astronomers now believe that dark energy, whatever it is, has been present since the time of the Big Bang, but in the much denser early universe, gravity was the dominant force, acting as a cosmic brake on the expansion. Then, about six billion years ago, the universe had thinned out enough for dark energy to become the dominant force, reversing the deceleration.

When HST was first launched, black holes with the mass of a few suns were known to exist. They were the end result of supernova explosions in massive stars that produced compact, collapsed cores with gravitational fields so intense that light itself could not escape its grip. Yet, data confirming the existence of supermassive black holes was tenuous and subject to intense debate. It wasn't because astronomers did not know how to look. While black holes are, by definition, not directly observable, astronomers have long known that any in-falling gas would be accelerated to extreme velocities, releasing high energy X-rays and gamma rays that could in theory, be seen. Unfortunately, using instruments to peer through Earth's atmosphere was a difficult challenge, and clear proof remained elusive.

Six months after the first shuttle servicing mission restored Hubble's vision, astronomers unveiled a pioneering observation of an active galaxy's core—M87—located some 50 million light years from the Earth. Hubble's Faint Object Spectrograph captured light from five positions around a rapidly spinning disk of gas at the core, allowing astronomers to measure its velocity.

According to Newton's law of gravity, the velocity of material in the disk depends directly on the mass of the material enclosed by it. When all the numbers were crunched, astronomers deduced that a black hole with a mass greater than 3 billion solar masses was responsible for the velocity of the material in the disk. The observation was hailed as 'textbook' proof that supermassive black holes really existed.

Subsequent observations by HST over the last two decades showed that not only do black holes exist, but that they are a fundamental part of every galaxy. Indeed, every galaxy that has been scrutinized in this way has revealed a massive black hole near their center. Subsequent work by HST showed that there is a strong correlation between the size of the galaxy and the mass of the supermassive black holes they contain. In other words, the mass of the black hole and the mass of the stellar bulge in any galaxy are tightly correlated. This means that the galaxy and the black hole at its center do not evolve independently of each other, but rather, they form and evolve together.

A sequence of images taken by HST over a period of 13 years revealed changes in a black-hole-powered jet of hot gas in the giant elliptical galaxy M87. The observations showed that the river of plasma, travelling at nearly the speed of light, may follow the spiral structure of the black hole's magnetic field, which astronomers think is coiled like a helix. The magnetic field is believed to arise from a spinning accretion disc of material around a black hole. Although the magnetic field cannot be seen, its presence is inferred by the confinement of the jet along a narrow cone emanating from the black hole. The visible portion of the jet extends for some 5000 light-years. M87 resides at the center of the neighboring Virgo cluster of roughly 2000 galaxies, located 50 million light-years away.

In July 1994, just seven months after the first shuttle servicing mission, fragments of a comet torn apart by Jupiter's enormous gravitational field slammed into the planet's atmosphere, blasting world-sized 'shiners' in the cloud tops that were easily visible to amateur and professional astronomers. The clearest, most spectacular views came from HST though—a powerful demonstration of the observatory's ability to provide flyby class views of the planets in Earth's solar system.

Hubble has been able to track Venusian clouds and dust storms on Mars, study the churning atmospheres of Jupiter and Saturn, monitor Saturn's majestic ring system and image spectacular auroral displays on both planets. Further afield, HST was used to keep tabs on Uranus and Neptune and their retinue of moons. More recently, Hubble has been used to map the moons of the dwarf planet, Pluto, and help find post flyby targets in the remote Kuiper Belt for NASA's Pluto-bound New Horizons spacecraft.

Getting spectacular images of Earth's neighbors was not a surprise, but actually imaging a planet orbiting another star, and spectroscopically examining the atmospheres of several other exoplanets, is considered a major achievement—and a major surprise.

Being able to characterize the atmospheres of these exoplanets is pretty remarkable, Riess said. "When Hubble was launched, we didn't even have evidence there were planets around other stars. Not only have those been found, Hubble has helped characterize them. It's truly remarkable!"

That milestone came in November 2008, when HST captured a visible light image of a planet orbiting the star Fomalhaut, some 25 light years away. With three times the mass of Jupiter, the planet appears as a faint point of light embedded in a vast ring of debris around the star. Getting an image of a planet that is one billion times fainter than its parent star was an achievement that stretched HST's power to the very limit.

More recently, Hubble has been used to make spectroscopic studies of giant exoplanets as they move around their host stars. By studying starlight that passes through the planetary atmospheres, researchers can tease out major constituents.

Box 2: How is the Hubble Space Telescope Pointed?

In a nutshell: via Newton's third law of motion, which states that for every action there is an equal and opposite reaction. To alter a body's orientation in space, a torque (turning force) must be applied around the body's center of gravity. There are several ways of achieving this, the most common being either ejection of mass (using gas jets, for example, in the same manner as a jet engine), or momentum transfer.

Gas jets are often used for fine adjustments in satellite orientation or of space vehicles such as the Space Shuttle. For astronomical observatories, however, jets have several drawbacks; their useful lifetimes (depending on the amount of gas that can be carried) are rather limited and they can potentially pollute the optical train. A better solution is to apply momentum transfer. This is done using a motorized flywheel mounted on the body of the spacecraft. By changing its speed, the spacecraft can rotate around its center of gravity, but in the opposite direction, thereby keeping its total angular momentum constant. Such flywheels are referred to simply as 'reaction wheels.' Three are needed for orienting the spacecraft in all directions, with a fourth usually added for redundancy.

The procedure for pointing the telescope in a new direction occurs in the following way. The rotation speed of the reaction wheel is increased or decreased accordingly. The telescope is then made to accelerate slowly, reaching a given speed and continues moving at this speed. When the desired new orientation is approached, the flywheel speed is brought back to its original value, bringing the telescope to a halt, precisely on target.

From an engineering perspective, the advantage of a space environment is that the vehicles can carry out these maneuvers with essentially zero friction, with no shafts, bearings or jittery motions. This allows them to reach remarkable levels of pointing accuracy and stability. Accuracy to the order of 0.1 arc second is trivial and it is possible to achieve a pointing accuracy some ten times better. HST, which actually holds the record in this domain, has a pointing stability of just 7 milli-arcseconds! To put this in perspective, an archer in New York would hit the bull's eye on a target located in Chicago! (Fig. 3.2).

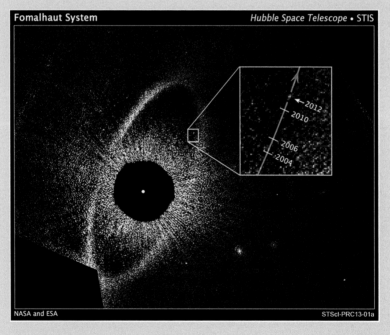

FIG. 3.2 Hubble images a spark—the planet known as Fomalhaut b—a billion times fainter than its parent star, Fomalhaut, some 25 light years away. Its orbital motion is also recorded (see inset image) (Image credit: NASA/STScI)

This new spectroscopic technology paved the way for more advanced surveys that will shortly be conducted by Hubble's successor, the James Webb Space Telescope (JWST), the topic of our final chapter.

Over the years, HST has taken some breathtakingly beautiful images of the cosmos, revealing its true majesty in the scheme of things. The most iconic of HST images is the "Pillars of Creation," part of Messier 16, located some 7000 light years away. Astronomers combined several Hubble exposures to assemble the wider view. The towering pillars are about 5 light-years tall. The dark, finger-like feature at the bottom right of Fig. 3.3 may be a smaller version

FIG. **3.3** 'The Pillars of Creation' in the Eagle Nebula, some 7000 light years away (Image credit: NASA/STScI)

of the giant pillars. The new image was taken with Hubble's versatile and sharp-eyed Wide Field Camera 3. The pillars are bathed in the blistering ultraviolet light from a grouping of young, massive stars located off the top of the image. Streamers of gas can be seen bleeding off the pillars as the intense radiation heats and evaporates it into space. Denser regions of the pillars are shadowing material beneath them from the powerful radiation.

Stars are being forged deep inside the pillars, which are made of cold hydrogen gas laced with dust. The pillars are part of a small region of the Eagle Nebula, a vast star-forming region 7000 light-years from Earth. The colors in the image highlight emission from several chemical elements. Oxygen emission is blue, sulfur is orange, and hydrogen and nitrogen show up in green (Fig. 3.4).

FIG. 3.4 The Tarantula Nebula as imaged by the Hubble Space Telescope (Image credit: NASA/STScI)

In the image above, a cluster of young stars sparkles with luminous energy and lights up a cavity in the roiling dust of the Tarantula Nebula. For Zoltan Levay, charged with bringing Hubble Space Telescope imagery to the public, the scene's dynamism is irresistible. In this image, some stars are being born while others are dying in the midst of cascades of gas and dust, constantly being churned by magnetic fields and powerful stellar winds.

The Tarantula Nebula has an apparent magnitude of 8, making it faintly visible in binoculars from a dark site. Considering its distance of about 49 kiloparsecs (160,000 light-years), this is an extremely luminous non-stellar object. Its luminosity is so great that if it were as close to Earth as the Orion Nebula, the Tarantula Nebula would cast shadows. In fact, it is the most active starburst region known in the Local Group of galaxies. It is also one of the largest such regions in the Local Group with an estimated diameter of 200 parsecs. The nebula resides on the leading edge of the Large Magellanic Cloud (LMC), where ram pressure is stripping and the potentially resulting compression of the interstellar medium is at a maximum (Fig. 3.5).

In the image above, Hubble's Wide Field Camera 3 looks through the Horsehead Nebula in a uniquely detailed infrared image. A classic target of astronomy, the nebula normally appears dark against a bright background, but Hubble penetrated the shroud of interstellar dust and gas. It's a hint of what to expect from NASA's planned infrared James Webb Space Telescope, due for launch in 2018.

The nebula is located just to the south of the star Alnitak, which is the star farthest east on Orion's Belt, and is part of the much larger Orion Molecular Cloud Complex. The nebula was first recorded in 1888 by Scottish astronomer Williamina Fleming on photographic plate B2312 taken at Harvard College Observatory. The Horsehead Nebula is approximately 1500 light years from Earth. It is one of the most identifiable nebulae because of the shape of its swirling cloud of dark dust and gases, which bears some resemblance to a horse's head when viewed from Earth (Fig. 3.6).

Described as a "million-second-long exposure", the Hubble Ultra Deep Field (HUDF), shown above, was taken during 400 orbits of the Hubble Space Telescope. HST made 800 exposures

Fig. **3.5** Hubble peers deep into the Horsehead Nebula in the constellation of Orion (Image credit: NASA/STScI)

over a four month period between September 24, 2003 and January 16, 2004. The average exposure time per image was 21 minutes. Hubble detected objects as faint as 30th magnitude. Even though the field of view was equivalent to looking at an area of sky visible through a straw some 8 feet long, it contains an estimated 10,000 galaxies, but the field of view is so empty that it contains only about seven stars from the Milky Way galaxy. It was taken by the ACS and NICMOS cameras (Fig. 3.7).

In the constellation Andromeda, which is 300 million light-years away, the two galaxies known as Arp 273 have interplaying

FIG. 3.6 The Hubble Ultra Deep Field showing galaxies billions of years into the past (Image credit: NASA/STScI)

gravitational forces. As if in a dance, the galaxies will orbit around each other for eons before finally coming together (Fig. 3.8).

The spectacular image above shows the spiral Sombrero Galaxy (Messier 104) seen almost edge on from Earth. The galaxy has a diameter of approximately 50,000 light-years, 30 % the size of the Milky Way. It has a bright nucleus and an unusually large central bulge along with a prominent dust lane in its inclined disk. The dark dust lane and the bulge give this galaxy the appearance of a Mexican sombrero. Astronomers initially thought that the halo was small and light, indicative of a spiral galaxy, but the Spitzer infrared telescope (discussed in the next chapter) showed that the halo around the Sombrero Galaxy is larger and more massive than previously thought. This is suggestive of a giant elliptical galaxy. The galaxy has an apparent magnitude of +9.0, making it easily visible by amateur telescopes and it is considered by some

Fɪɢ. 3.7 The interacting galaxies collectively known as Arp 273, captured by HST, lie 300 million light years away (Image credit: NASA/STScI)

Fɪɢ. 3.8 The majestic edge-on galaxy affectionately known as the Sombrero Galaxy, as captured by HST (Image credit: NASA/STScI)

authors to be the brightest galaxy within a 10 megaparsecs radius of the Milky Way. The large bulge, the central supermassive black hole and the dust lane all continue to attract the attention of professional astronomers.

These are just a few examples from the extensive library of Hubble images that have greatly enriched our knowledge of the universe, revealing its grandeur and unparalleled beauty. Though it was in danger of being mothballed, HST has redeemed itself, and will go down in history as arguably the most famous space telescope of all time. In the next chapter, we'll take a closer look at the infrared universe and the space telescopes that revolutionized our understanding of invisible realms of the cosmos.

4. The Infrared Universe

The year was 1800. Sir William Herschel used a prism to refract light from the sun and detected heat rays beyond the red part of the spectrum, recording an increase in temperature with the use a thermometer. He was surprised at the result and called them "Calorific Rays". The modern term "Infrared" (IR) did not appear until the late nineteenth century.

Infrared is usually associated with thermal radiation, which is composed of rays that are emitted by all warm bodies. It is the basis behind "night vision" and thermal imaging technology used in hospitals. The ability to image in the IR region of the spectrum has been used by fire fighters for many years. Because these rays are longer than visible radiations, they can help penetrate smoke-filled rooms, allowing people to be more easily identified in rescue missions. In astronomy, IR has helped us peer through dusty galaxies, glimpse protoplanetary disks around young stars and witness the birth of stars inside their dust-laden cocoons.

Infrared spans a waveband between 700 nm to about 1 millimeter. Optical telescopes are capable of detecting IR radiation without any significant modifications to their design but the Earth's environment greatly reduces the efficiency of capturing and focusing these radiations. Two basic problems attend ground-based IR astronomy, and they are not easily resolved. For one thing, the Earth's atmosphere generally looks opaque using infrared radiation even though the peak of the black body curve (Chap. 1) for both the sky and ground-based objects, such as telescopes at ambient temperature, lies in the infrared band. Furthermore, water vapor, in particular, is a potent absorber of IR radiation, and as a result, ground-based telescopes must be placed at very high altitudes to efficiently detect these radiations from celestial sources. Any ground-based IR telescopes must be cooled to the lowest temperature possible to get the best IR images. By placing these telescopes in space, astronomers can greatly improve their efficiency, allowing mankind to peer deeper into the mysteries of creation.

© Springer International Publishing Switzerland 2017
N. English, *Space Telescopes*, Astronomers' Universe,
DOI 10.1007/978-3-319-27814-8_4

The American astronomers Gerald Neugebauer and Robert Leighton conducted the first pioneering IR surveys employing a simple telescope with a 1.5-meter aperture and equipped with a lead sulfide detector (a photodiode) cell cooled using liquid nitrogen. The instruments were optimized to monitor radiation with a wavelength of 2.2 microns and their ground-breaking work produced the first IR survey—the so-called Two Micron Survey, which was first published in 1979. In this study, they identified some 5612 IR sources, most of which turned out to be highly evolved red giant and supergiant stars.

Their work was complemented and extended by a series of nine rockets, each carrying a 16.5 cm diameter IR telescope, under a project code named HISTAR and directed by the US Air Force Cambridge Research Laboratory. Each of these instruments were actively cooled using liquid helium and carried out surveys in the 10–20 micron range as well as more limited surveys at 4 and 27 microns. Remarkably, while the total observing time of these telescopes was only half an hour or so, their observations yielded over 2000 IR sources, mostly from stars, nebulae and interstellar clouds radiating at blackbody temperatures from about 100 to 1500 K (Fig. 4.1).

It was a team of Dutch scientists and engineers who were the first to bring the idea of a dedicated IR space telescope to the table, but they soon found that its construction was too costly for one nation to pick up the tab. It was therefore incumbent upon other

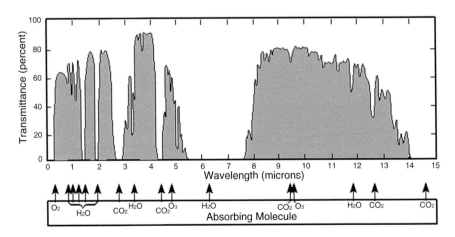

FIG. 4.1 Showing the opaqueness of the Earth's atmosphere to infrared radiation (Image credit: NASA)

nations to make this a reality. When they brought their ideas to NASA, the organization expressed a genuine interest in the project, also soliciting the expertise of scientists and engineers based in the UK. In 1973, a formal agreement was established for a collaborative project between these nations to build a dedicated orbiting Infrared Astronomical Satellite (IRAS). NASA was to build the telescope, its state-of-the art cooling system, as well as the launcher. Meanwhile the Dutch team would design and construct the spacecraft bus, while UK engineers would be responsible for building the ground station where all the data analysis would subsequently be undertaken.

That said, making IRAS a technological reality was easier said than done. It would involve a decade-long collaborative effort between the three countries. Almost as soon as work began, problems began to surface. For one thing, the construction of the mirror, which was commissioned out to Perkin Elmer, was stalled after it was discovered that it had developed a large scratch. Moreover, several faults were found with the electronic detectors and cooling system, which resulted in a revision of the proposed mission lifetime from an estimated 460 days to just 220 days. To add insult to injury, more technical glitches were uncovered just a few months before its proposed launch date, which collectively served to make officials rethink whether IRAS should be mothballed or at least postponed until the whole project could be executed more thoroughly. Meanwhile, the cost of IRAS spiraled from an estimated $37 million to over $92 million, yet NASA still gave the all-clear to launch the satellite at the beginning of 1983.

The IRAS payload consisted of a 57 cm f/9.6 Ritchey Chretien reflecting telescope with a true field about 30 arc minutes (half an angular degree or about the size of the full moon as it appears in our sky). The mirror was cooled to a temperature below 10 K using 72 kilograms of superfluid helium contained in a large vacuum flask. An array of 62 rectangular semiconductor detectors in the telescope's focal plane was kept just 2.5 degrees above absolute zero (2.5 K). The detectors, which were designed to cover broad wavebands centered on 12, 25, 60 and 100 microns, were arranged in such a way that each source crossing the field of view would generally be monitored by at least two detectors in each of these wavebands. Such an approach would greatly reduce the number of false or 'spurious' signals.

IRAS' main mission was to obtain an all-sky survey at wavelengths from 12 to 100 microns at an angular resolution of 4 arc minutes, as well to conduct detailed observations of selected targets using its on-board Low Resolution Spectrometer (LRS) and the Chopped Photometric Channel (CPC). The LRS was designed to measure the spectra of bright sources over a wavelength range of 8 to 23 microns. The CPC complimented this work by mapping extended IR targets at a resolution of just 1 arc minute over a waveband spanning 41–63 microns. Final pinpointing of the target was done by small optical detectors which could increase the resolution to 20 arc seconds.

Weighing in at one ton, IRAS was launched into space on January 26, 1983 from the Western Test Range in California into a 900 kilometer altitude polar orbit. Within just six months, the orbiting infrared observatory completed its first all-sky survey. Because of its low polar orbit, communications with the ground based station at the Rutherford Appleton Laboratory (RAL) in England were curtailed to brief intervals twice a day. As a result, the data from IRAS was updated every 12 hours. In addition IRAS had a limited data storage capacity on its onboard tape recorders and transmitted to the ground station at a rate of 1 Mbps every 12 hours. After preliminary analysis of the data by scientists at the Rutherford Appleton Laboratory, the same IRAS images underwent more refined processing at the Jet Propulsion Laboratory (JPL), California, USA.

After launch, IRAS encountered some teething problems. Specifically, its primitive onboard computer went into 'failsafe mode' owing to spurious signals from a solar source, which had to be corrected by a team of programmers at JPL. After 300 days in orbit, IRAS ran out of its liquid helium coolant and the mission came to an end on November 22 1983. Yet, despite this rather short-lived mission, IRAS was deemed a great success by mission controllers. It had managed to survey 95 % of the sky at least four times over. The data unveiled by IRAS was published the following year and contained data on 245,000 separate sources including solar system, galactic and extra galactic objects.

One of the most intriguing discoveries made by IRAS was the discovery of so-called 'interstellar cirrus,' enormous clouds of dust hanging out between the stars and radiating at 35 K, corresponding to a peak wavelength of about 100 microns. Such dust was first brought to the attention of the astronomical community during the 1970s when the American astronomer, Allan Sandage, uncovered some evidence of thick clouds of dust obscuring background starlight, especially at high galactic latitudes. After analyzing images made by IRAS, Frank Low and his colleagues at the University of Arizona detected streaky patches of cool dust strewn randomly all over the sky. This dust most likely consists of small carbon particles that adhere to hydrocarbon ice grains. This interstellar cirrus was found to be particularly concentrated within regions of neutral hydrogen in the Milky Way, indicating that the majority of it resides within our own galaxy.

When IRAS turned its instruments on the bright star Vega, it discovered an extensive disk of dust in its immediate vicinity, radiating in the infrared at a temperature of 80 K. Further analysis of this dust disk showed that the particles were most likely about 1 mm in size. With an estimated age of only a few hundred million years, Vega's dust disk was the first evidence that it may be in the process of forming planets. Observations of the star Beta Pictoris showed a similar infrared excess indicative of a protoplanetary disk about 800 astronomical units in diameter.

Indeed, IRAS went on to show that many stars, particularly highly evolved ones, are surrounded by cool dust and that this is part of the general mechanism of mass loss in older stars. What is more, IRAS was able to identify and peer deeper into Bok globules, almost universally accepted to be the birthplace of stars. Such observations greatly aided astronomers trying to solve the riddle of how stars form from matter in the interstellar medium.

Further afield, IRAS showed that many galaxies were extremely strong radiators in the infrared region of the spectrum. Indeed, some so-called 'peculiar' and older elliptical galaxies were shown to be 80 times brighter in the infrared than at visible wavelengths! This is thought to reflect the large number of red giant stars within these galaxies. Other infrared galaxies were shown to have undergone recent mergers with other galaxies in the near past, creating new waves of star formation.

In the years after the cessation of IRAS, astronomers built a number of ground based telescopes that could better image in the infrared. But because of the great advantages of an orbiting IR observatory, scientists were soon looking for a replacement of the highly successful IRAS mission. This led to the design of the Spitzer Space Telescope. The telescope was named after the great American astronomer, Lyman Spitzer, Jr. (1914–1997), who was one of the twentieth century's great scientists. A renowned astrophysicist, he made major contributions in the areas of stellar dynamics, plasma physics, thermonuclear fusion, and space astronomy. Lyman Spitzer, Jr. was the first person to propose the idea of placing a large telescope in space, and was the driving force behind the development of the Hubble Space Telescope.

At a cost of US $800 million it was launched from Cape Canaveral Air Force Station, on a Delta II 7920H ELV rocket, on Monday, 25 August 2003. Compared to any previously designed infrared telescope, the Spitzer Space Telescope was a bona fide marvel of high technology, incorporating many innovations never before used on a space mission. Some 4 meters (13 feet) tall, the telescope only weighed 865 kilograms (1906 pounds). Since Spitzer was created to detect infrared radiation, or heat—its detectors and telescope had to be cooled to only about 5 degrees above absolute zero (–450 degrees Fahrenheit, or –268 degrees Celsius). This ensured that the heat generated by the observatory could not interfere with its observations of relatively cold cosmic objects. Intriguingly, while parts of Spitzer had to be kept cold to function properly, other on-board electronic devices required near room temperature to function properly. In order to achieve this balance of warm and cold, the telescope had to be compartmentalized into two components:

The Cryogenic Telescope Assembly, which incorporated Spitzer's cold components, the most important of which included a 0.85-meter telescope and three scientific instruments.

The Spacecraft—which contained components that could only function at higher temperatures, including the solar panels, the various telescope controls as well as the tools employed to communicate scientific information back to scientists on Earth (Fig. 4.2).

The primary instrument package (including the telescope and cryogenic chamber) was designed and put together by Ball Aerospace & Technologies Corp., in Boulder, Colorado. The individual instruments used by Spitzer were constructed piecemeal by

FIG. **4.2** Artist's conception of the Spitzer Space Telescope exploring the infrared universe (Image credit: www.spitzer.caltech.edu)

collaborative work conducted between industrial, government and industrial institutions, including Cornell, the Smithsonian Astrophysical Observatory and the University of Arizona, Ball Aerospace, and engineers based at the Goddard Spaceflight Center. The infrared detectors were the brainchild of Raytheon, a company based in Goleta, California. Raytheon engineers coupled the properties of indium antimonide and a doped silicon detector in bringing the infrared detectors to life. These detectors were an enormous advance over the IR detectors used in the 1980s, being 100 times more sensitive. The spacecraft itself was built by Lockheed Martin. The entire mission was to be operated and directed by astronomers at the Jet Propulsion Laboratory and the Spitzer Science Center, located on the Caltech campus of the University of California, Pasadena.

Spitzer was placed in a rather unusual orbit, which was heliocentric instead of geocentric. This meant that the spacecraft railed and drifted away from Earth's orbit at a rate of approximately 0.1

FIG. **4.3** A glimpse of the infrared cirrus—delicate gossamer threads wrapping their way round the stars of the Pleiades (Image credit: NASA)

astronomical units per year. Spitzer's primary mirror was 85 centimeters (33 in) in diameter with a focal ratio of f/12 and fashioned from beryllium, a metal which is of comparatively low density and is thus lightweight and also withstands cryogenic temperatures better than many other metals. The optics were cooled to just 5.5 degrees above absolute zero using 360 liters of liquid helium. The satellite was fitted with three instruments that allowed Spitzer to perform both imaging and photometry at wavelengths from 3 to 180 microns, spectroscopy from 5 to 40 microns, and spectrophotometry from 5 to 100 microns (Fig. 4.3).

The optics of the Spitzer telescope were diffraction limited, that is, it produces images that are accurate to within one quarter wave of green light (550 nm), and in this segment of the infrared spectrum, this amounts to a resolving power of just 6.5 microns. At its launch into space in August 2003, Spitzer was designed to work for a minimum of about 30 months but it would ultimately give research astronomers bountiful views of the infrared universe for an impressive five and a half years.

On May 15, 2009, having depleted all its liquid helium coolant, Spitzer formally ended its 'cool' exploration phase, but it then began a new life as a purely optical telescope, where it is anticipated to work until at least 2020.

Spitzer's highly sensitive instruments allowed scientists to penetrate into hidden realms of the cosmos which were almost entirely hidden from the sensitivities of telescopes working over visible wavelengths. Spitzer's large mirror could peer deep inside dusty stellar nurseries, the inner sanctum of galaxies, and even newly forming planetary systems. Spitzer's infrared eyes also allowed astronomers to see very cool objects in space, including failed stars—the teeming multitudes of brown (L) dwarves—extra-solar planets and giant molecular clouds. Its precision infrared spectrometer would sniff out the intricate chemistry of individual stars, as well as the molecules that abide in the vast spaces between the stars—what astronomers call the interstellar medium—searching for the tell-tale signs of chemical elements and organic compounds.

Stars are born deep inside vast inkblots in space, from condensing clouds of gas and dust. Technically these embryonic suns do not visibly "light-up" until they've acquired enough mass and gravity to ignite thermonuclear fusion in their cores. As the collapsing cloud becomes increasingly massive, its gravitational field becomes stronger. Because these structures initially develop behind a thick veil of cosmic dust, optical telescopes are all but blind to the first chapters in the allegory of star formation, but not NASA's Spitzer Space Telescope. Like a giant night vision 'scope, Spitzer's heat-sensitive infrared eyes were able to capture the first violent portraits of stars in the midst of their formation. Many of the images Spitzer captured have been processed and color-coded for greater clarity. In the images that follow, regions that show up as red are cooler than those that are green, while blue represents the hottest temperatures (Fig. 4.4).

With Spitzer, "it's like having an ultrasound for stars," said project scientist, Dr. Jeonghee Rho. "We can see into dust cocoons and visualize how many embryos are in each of them" (Fig. 4.5).

As discussed earlier in the book, stars are celestial chemical factories, living out their entire lives fusing hydrogen and helium atoms into a suite of heavier elements. In death, extremely massive stars explode in cataclysmic supernova events, blasting their chemical creations into space, and in so doing, seeding the universe for a new generation of stars to form. Meanwhile, medium mass stars like our Sun balloon in size to become red giants before shedding their outer layers, in much the same way snakes shed their skin, sending newly-formed elements and molecules into the great dark of interstellar space. By studying geriatric stars, Spitzer provided valuable new insights into the composition of their dusty death

Fig. 4.4 Spitzer images protostars in the making (Image credit: NASA)

Fig. 4.5 Another infrared image from Spitzer showing embryonic stars within the Triffid Nebula (Image credit: NASA)

throes, and clues as to how they died. These studies also provide crucial evidence that may help astronomers solve the longstanding mystery of where the dust in our very young universe came from.

In peering deep inside star forming regions, Spitzer has reminded astronomers that they are perilous places for planets and life. In the image shown below, Spitzer has used its giant infrared eye to catch young, hot stars stripping planet-forming material off of younger stars. The powerful stellar winds whipped up by these newly formed suns can strip planets of their atmospheres, making them desolate places with no chance for life. We should thank our lucky stars that our Sun escaped its birth cluster before any such destruction occurred to its retinue of planets, including the Earth (Figs. 4.6, 4.7 and 4.8).

Infrared emission from most galaxies comes primarily from three sources: stars, interstellar gas and dust. With NASA's Spitzer Space Telescope, astronomers can see which galaxies are furiously

Fɪɢ. 4.6 Star forming regions like this one captured by Spitzer can be perilous places for planets (Image credit: NASA)

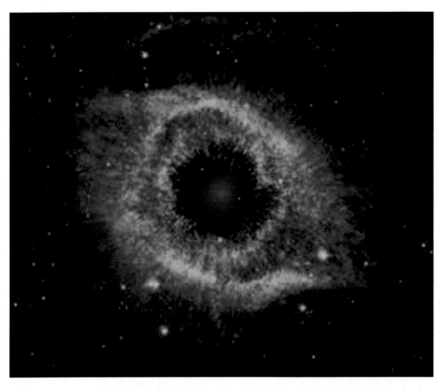

FIG. 4.7 Spitzer images the 'Eye of God'. The Helix Nebula is a highly evolved middle mass star ejecting its outer layers to the cold dark of interstellar space (Image credit: NASA)

FIG. 4.8 Spitzer infrared image of the Andromeda Galaxy (M31) showing its dusty spiral arms (Image credit: NASA)

forming stars, pinpoint the location of stellar nurseries, and identify the factors which trigger these stellar baby booms.

In looking at our own Milky Way, Spitzer has provided astronomers valuable new insights into the structure of our home galaxy, by showing them where all the new stars are forming.

These studies have shown that star formation is almost exclusively occurring in the densest regions of their spiral arms. Elliptical galaxies, in sharp contrast, are almost completely devoid of star forming regions, their constituent stars being predominantly made up from highly evolved red giants and copious amounts of dust.

Spitzer also played a significant role in helping astronomers understand Ultra Luminous Infrared Galaxies, or ULIRGs. These galaxies emit more than 90 % of their light in the infrared, and are primarily found in the distant universe (high red shifts). With Spitzer's giant infrared eye, astronomers were able to determine whether the source the ULIRG's infrared glow was emanating from extreme star-formation or an active central supermassive black hole, or both (Fig. 4.9).

The Spitzer telescope was able to show that the bulk of this infrared 'bleeding' was due to spectacular bouts of star formation. Indeed, the peak epoch in the universe's propensity to create new stars happened many billions of years ago: 90 % of all stars that will ever exist in our universe have already been formed!

FIG. 4.9 This Spitzer image of the Whirlpool Galaxy (M51) shows that the galaxy is rich in dust, with regions of vigorous star formation. The blue galaxy at the top is considerably older than M51 (Image credit: NASA)

The bright and massive stars that dominate our night sky are actually a minority in the cosmos. Most stars are far too puny to even ignite thermonuclear fusion in their cores, thus they never visibly light up. Instead they emit primarily infrared light, which the Spitzer can easily detect. Astronomers refer to these "failed stars" as brown dwarfs.

Some astronomers claim that brown dwarfs may prove to be extremely important contributors to the mystery of "dark matter." Galaxies are heavier than they look, and scientists have adopted the term "dark matter" to loosely define all the material in the universe that can't be seen. As well as this ordinary, non-luminous matter, a large fraction of dark matter is very probably made up of exotic materials, different from the ordinary particles that make up the familiar world around us. Future research may shed significantly more light on the nature of this exotic material.

Spitzer's supersensitive infrared eyes have discovered thousands of brown dwarfs, and, intriguingly, have unearthed evidence that many of these failed stars may even harbor planets (Fig. 4.10).

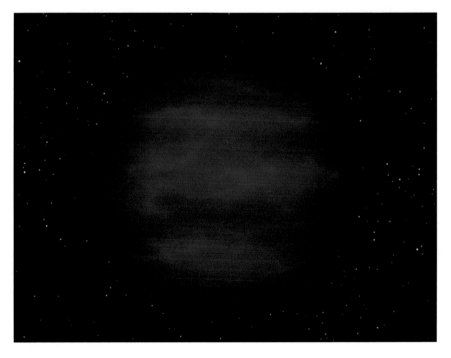

FIG. **4.10** Artist's conception of a Brown Dwarf (Image credit: NASA)

In a later chapter, we shall explore the prospects of the most ambitious infrared telescope ever conceived of by mankind; a telescope that, in comparison, will make Spitzer's achievements pale into insignificance. This is the extraordinary James Webb Space Telescope (JWST) scheduled for launch in 2018.

In the meantime though, let us continue our exploration of the cosmos as revealed by other kinds of telescopes probing entirely different provinces of the electromagnetic spectrum. In so doing, we turn from the low temperature universe to the high temperature universe; from the infrared to gamma rays.

5. The Gamma Ray Universe

Thus far, we have confined our discussion to space telescopes capable of detecting visible and infrared sources. But the hottest and most violent events in the cosmos produce the highest energy EM radiation of all; gamma rays. These rays are deadly to all known life forms but, as we shall see, their study has unveiled a high energy cosmos that adds a whole new dimension to our understanding of the universe and how fortunate we are to live in a relatively safe place far from any gamma ray sources.

Gamma rays are photons just like visible light or infrared radiation. But, owing to the fact that they are hundreds or even thousands of times more energetic than X-rays, they travel through space essentially unimpeded. Practically the entire universe is transparent to gamma rays, so they may appear to be the perfect tools to allow astronomers to 'see forever', as it were. Fortunately for us however, after surviving a journey of millions or billions of light years, cosmic gamma rays are arrested just a few kilometers above where earth-based telescopes are located. When these ultra-high-energy photons buffet our atmosphere, they collide with atoms and molecules, which in turn produce showers of other kinds of subatomic particles. In so doing, they could be said to sacrifice their original identity, which puts limitations on their ultimate usefulness for doing astronomical research.

The first forays into the brave new world of gamma ray astronomy came in the early 1960s. Scientists and engineers first opened the door of the gamma ray universe when they discovered a weak but diffuse glow of gamma radiation bathing the entire sky, using a primitive detector flown aboard an Explorer satellite. During the next 20 years, American, Soviet and European spacecraft showed that virtually all places in the cosmos where violent, energetic processes occurred, gamma rays were almost invariably produced.

Our own star, the Sun, emits gamma radiation originating from the magnetically active regions associated with solar flares. While usually manifesting themselves as radio pulsars, rapidly

© Springer International Publishing Switzerland 2017
N. English, *Space Telescopes*, Astronomers' Universe,
DOI 10.1007/978-3-319-27814-8_5

rotating neutron stars can also generate gamma ray sources. These high-energy emissions are produced via the complex interplay between strong gravitational fields, super-strong magnetic fields and the ultra-high surface temperatures they generate. Objects which may harbor supermassive black holes in their centers, like quasars and active galaxies, are known to produce unstable (or even erratic) gamma radiation.

All these celestial objects may produce not only a broad continuum of gamma radiation, but also highly specific emission lines with highly particular energies. As we have seen, such emission lines can unambiguously identify the chemical elements and the undergirding physical processes that gave rise to them and much in the same way the Fraunhofer absorption lines in optical spectra of stars can identify the atoms, ions and molecules in stellar atmospheres.

Because of their exceedingly high energies, gamma rays are much harder to produce than other forms of electromagnetic radiation. Not only that, they are harder to detect as well, requiring instruments that are technically complex and difficult to use. To get an idea of how difficult it is to do gamma ray astronomy, consider this: even though several small gamma ray telescopes were launched into space throughout the 1980s, they collectively only gathered about 200,000 gamma rays!

The first gamma-ray bursts were detected in the late 1960s by the U.S. Vela satellites, which were built to detect high energy radiation pulses emitted by enemy-launched nuclear weapons tested in space. The United States government believed that the USSR was carrying out secret nuclear tests even though they had signed the Nuclear Test Ban Treaty of 1963. On July 2 1967, both the Vela-3 and 4- satellites detected a burst of gamma radiation quite unlike, in character, any known nuclear weapon-induced flash. Uncertain what had happened, but not considering the matter particularly urgent, the team at the Los Alamos Scientific Laboratory, led by Ray Klebesadel, stored the data away for investigation at a later date. As additional Vela satellites were launched with better instruments, the Los Alamos team continued to find inexplicable gamma-ray bursts in their data. By analyzing the different arrival times of the bursts as detected by different satellites, the team was able to determine rough estimates for the sky positions of sixteen bursts and definitively rule out a terrestrial or solar origin. The discovery was declassified and published in 1973 as an Astrophysical Journal article entitled "Observations of Gamma-Ray Bursts of Cosmic Origin."

Designing gamma ray telescopes with enough resolution to pinpoint a source was a formidable design challenge for the earliest gamma ray detectors launched into the vacuum of outer space. The first space-based detection of gamma rays came from data collated by the American Explorer 11 satellite back in 1961. Six years later, gamma ray astronomy was placed on a much firmer footing, when OSO-3 identified the Milky Way galaxy as a strong source of gamma rays, especially along the galactic equator. That said, it is noteworthy that OSO-3 only detected 600 gamma ray photons during its 31 month long mission. But this was an era of rapidly advancing technology and subsequent missions greatly advanced our knowledge of the gamma ray universe.

During its seven month lifespan, the US-made SAS-2 satellite improved the gamma ray detection rate by another order of magnitude, recording up to 8000 photons with energies above 25 MeV. And despite its limited angular resolution (~1.5 degrees) SAS-2 was able to show that the Galactic center, as well as the Vela and Crab pulsars, was a highly active gamma ray source. The mission was cut short due to a mechanical fault on this early gamma ray telescope, which meant that only half the sky was surveyed. Despite these modest advances, interest in gamma ray bursts among high energy astrophysicists was stoked so much so that they began to think about strategies to overcome their pointing accuracy limitations. It wasn't long before a simple solution was found. By employing two telescopes aimed at an identical source, the object could be pinpointed by means of simple triangulation.

The European Space Research Organization (ESRO) formally expressed its interest to enter the brave new world of gamma ray astronomy, when it launched its very own gamma ray telescope back in 1969. Known as Cos-B, the orbiting satellite covered an energy range from 30 to 5000 MeV. Remarkably, this satellite continued to be operational until April 25, 1982, far longer than anyone thought possible. In total, Cos-B, recorded 100,000 gamma ray photons from all across the sky. Its great longevity demonstrated that repeated gamma ray observations could be made, identifying 25 new sources in the process (Fig. 5.1).

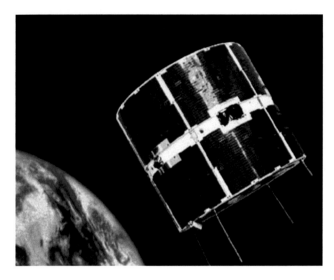

Fig. 5.1 Artist's conception of ESRO's Cos-B gamma ray satellite (Image credit: NASA)

The next wave of advances in gamma ray astronomy took place during the late 1970s when NASA launched their third High Energy Astronomy Observatory (HEAO). Called HEAO-3, the spacecraft was launched from Cape Canaveral on September 20, 1979. Although the gamma ray telescope on-board the spacecraft produced its fair share of spurious results, it was the first of its kind to identify Cygnus X-1 as a powerful source of gamma rays. What's more, HEAO-3 showed that its gamma ray emission was far from continuous. Cygnus X-1 is now believed to be a black hole that periodically accretes matter by snacking on any matter that delves too close to its event horizon.

In 1977, NASA's High Energy Astronomy Observatory program formally announced plans to build a great observatory dedicated to gamma-ray astronomy. The Compton Gamma-Ray Observatory (CGRO), named after the American physicist, Arthur Holly Compton (1982–1962), was designed to take advantage of the major advances in detector technology that took place during the 1980s. The observatory was finally launched into its 450 kilometer altitude orbit in 1991 at a cost of $617 million. Weighing in at 17 tons, CGRO became the largest unmanned civilian satellite carried into Earth orbit. GCRO had four major instruments on-board, all of which were at least 10 times more capable than the previous generation of gamma ray telescopes and considerably more sensitive, which in turn, greatly improved both the spatial and temporal resolution of subsequent gamma ray observations. Let's take a closer look at the instruments carried by this state-of-the-art gamma ray observatory.

The Burst and Transient Source Experiment (BATSE) was designed to detect ultra-short gamma ray bursts in the energy range 0.03 to 1.9 MeV. BATSE consisted of eight identical sensors positioned at each of the corners of the spacecraft, which provided all-sky coverage. While the pointing accuracy of the individual BATSE sensors was about two angular degrees (four times the apparent diameter of the full Moon), if two or more sensors detected the same gamma ray burst, triangulation could be used to greatly improve the resolution of the images down to one arc minute (that's about 1/30th of the size of the full Moon). In addition, an array of secondary detectors covering a gamma ray energy range from 0.015 to 110 MeV were used to monitor the energy of gamma ray bursts, although with considerably less sensitivity than the former. But once a burst was recorded and pinpointed, BATSE would activate its three other instruments on board the spacecraft.

The Oriented Scintillation Spectrometer Experiment (OSSE), built by the Naval Research Laboratory, was designed to detect gamma rays entering the field of view of any of the four detector modules. These could be pointed independently of each other, and were effective in the 0.05 to 10 MeV energy range. These four detectors moved back and forth over a large arc some 192 degrees in size, monitoring the background sky for the tell-tale signs of high energy radiation bursts.

GGRO also carried the so called COMPton TELescope (COMPTEL) instrument, which was deigned to image gamma ray sources over an energy range of 1 to 30 MeV, with a wide field of view covering about 60 degrees of sky with a pointing accuracy of 30 arc minutes. The last instrument on board the Compton satellite, the so-called Energetic Gamma Ray Experiment Telescope (EGRET) was designed to detect the highest energy gamma rays in the range 20 to 30,000 MeV, but over a narrower field of view than COMPTELL. EGRET could detect the strongest sources to an accuracy of just 5 arc minutes (Fig. 5.2).

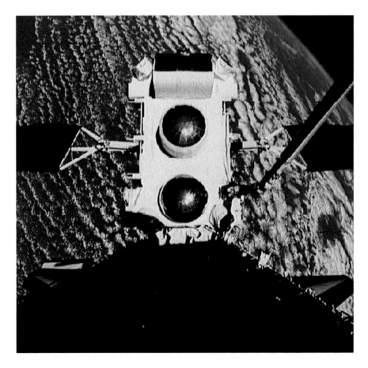

Fig. 5.2 The Compton satellite during its deployment from the Space Shuttle into earth orbit (Image credit: NASA)

The Sun smiled on the Compton observatory shortly after launch. In late May 1991, both OSSE and BATSE instruments detected copious gamma ray emission from an active region of the solar disk. By June 7 of the same year, the gamma ray outbursts had become so intense that mission controllers were forced to suspend routine operations and declare the Sun a novel "target of opportunity". When the spacecraft was pointed toward our star, the instrument's gamma ray counting rates were literally off the scale. That said, pointing the spacecraft at the Sun resulted in dangerous increases in temperature of the on-board electronics, causing its sensitive circuitry to struggle to cope with the added thermal strains.

Nonetheless, the Compton orbiting observatory managed to record a broad continuum of high energy emissions, as well as spectral lines deriving from oxygen, carbon, nitrogen and a variety of other atomic nuclei. The source of these gamma rays were thought to derive from nuclear reactions on the Sun's surface, triggered by the impact of charged particles which were accelerated to very high energies by the powerful magnetic fields associated with solar flares. Together, the four instruments on board Compton were able to collect solar photons with energies in the range from 20,000 to 1GeV, and representing a dynamic range of 50,000. To gain some appreciation of this range, compare this with visible light, where violet photons are only twice as energetic as their red counterparts.

The solar flaring which occurred in June 1991 provided an excellent opportunity to demonstrate Compton's formidable abilities. Barely two months into a very complex mission, spacecraft controllers responded very rapidly to this new "target of opportunity", dedicating a sizeable chunk of the spacecraft's observing time available to monitoring the Sun. Ingenious on-board signal processing enabled the instruments to communicate with each other and to change their operating modes automatically and appropriately as the Sun went from a state of quiescence back to high flare activity. This remarkable degree of synchronicity enabled the four experiments to return data that amounted to much more than the sum of its parts.

It is estimated that our Galaxy contains between 100 million to 1 billion stars that ended their evolutionary journey by exploding as supernovae. In the aftermath of these violent stellar explosions, compact neutron stars were left behind with masses about equal to that of our sun but having diameters of only 10 kilometers or so. Although these stellar remnants were squarely predicted from theoretical calculations, their direct detection was an altogether more difficult prospect.

Having no reserves of nuclear fuel to burn, a neutron star cannot emit visible light in the same way that a normal star can. But that does not mean that it dies away quietly. Astronomers now know that it can be even more active than its progenitor star was. A rapidly spinning neutron star with a strong magnetic field will accelerate charged particles in its vicinity, causing it to emit electromagnetic radiation. Most have lower energies than visible light photons, so the resulting emission falls largely in the radio spectrum, but gamma ray pulses have also been identified.

The pulsar at the heart of the Crab Nebula (Messier 1) was one of only two known to emit gamma rays before the inception of the Compton mission (the other being the Vela pulsar). This rapidly spinning neutron star spins on its axis about 30 times each second, emitting a strong main pulse followed by a weaker inter-pulse. This pulsation scheme is caused by a cone shaped 'search-light' beam of photons that sweeps across the plane of the Earth's orbit with each revolution it completes. What's more, these pulses have been observed right up through the highest gamma ray ener-gies detectable.

Any pulsar emitting gamma rays must be capable of transfer-ring large amounts of energy to subatomic particles as they are accelerated along super-strong magnetic fields. Although there were many ideas about how such a transfer of energy might occur, the exact mechanism remained a mystery. Not surprisingly, one of the Compton's observatory's prime objectives was to find more gamma-ray pulsars among the many other known radio pulsars in the Galaxy. Astronomers could work hand-in-hand with nature to determine the pulsation periods of these extraordinarily accurate timing devices. By monitoring pulsed gamma ray emissions with just the right periodicity, the GCRO would be able to hone in on all known radio pulsars, one at a time, to see whether or not they were 'broadcasting' at gamma ray energies too.

In January 1992, the team assigned to BATSE announced the detection of a third gamma ray pulsar, this time in the constellation Circinus. The period of its gamma ray pulses matched its radio-wave counterpart of seven rotations per second. More success came fast on the heels of the Circinus pulsar, when scientists announced that the EGRET had unveiled a fourth gamma ray pulsar, in Scorpius, which emitted photons with energies exceeding 100 mil-lion eV. Unlike the two earlier examples, the new Scorpius gamma ray pulsars had an emission period somewhat slower. At first, this was considered to be rather mysterious. But when more light was shed on the data further study, and after additional gamma ray pul-sars were unearthed, the GCRO data provided invaluable new insights into the mechanisms driving particle acceleration in the intense gravitational fields found in the vicinity of neutron stars.

Gamma ray bursts were once one of the most puzzling of astrophysical phenomena. About once every day, without warning, a flash of gamma radiation typically lasting 10 seconds arrives from some random point in the sky. Once the event is over, the source again becomes completely invisible at all wavelengths. Understanding these cosmic gamma-ray bursts was one of the

GCRO's main mission goals, and all four of its instruments could record them over a wide range of energies.

Astronomers were stabbing in the dark for quite some time as to the precise nature of the gamma-ray bursters, having to rely on clues derived from light curves and a close examination of the spectral lines. Rapid brightening indicated that small objects were involved, since such rapid changes could not have occurred in less time it took for the light to travel across the source. The brightness of some bursts fluctuated over fractions of a millisecond, strongly suggestive that the source diameter was of the order of about 60 kilometers. What is more, the emission and absorption lines observed in the spectra of some bursts showed that the objects responsible for them were unusually dense and possessing intense magnetic fields.

Because neutron stars are so tiny—at least in stellar terms—super dense and strongly magnetic, they appeared to fit the buster puzzle nicely. But ever since they were first unveiled in the mid-1970s, gamma ray bursts have been shown to come from all points in the sky, uniformly, and in about equal numbers, whereas neutron stars were shown to be most highly concentrated near the Galactic disk. Before GCRO's launch, the solution of this mystery seemed to lie with detector sensitivity. For example, when we peer up from a light-polluted night sky, we observe stars scattered equally over the entire sky, because we see only the brightest and nearest members. But from a dark, country sky, many more stars of various degrees of glory and the Milky Way itself become clearly visible. Thus, observing from a new location increases our sensitivity, so we see farther out into the Galaxy.

While pre-Compton burst experiments were small and had limited sensitivity, the BATSE detectors were 20 times more powerful than anything flown before. As a result, most astronomers expected them to 'see' beyond the solar neighborhood and deep into the galactic disk. With these greatly enhanced detection capabilities, gamma-ray bursts were soon found to concentrate along the Milky Way and soon the last piece of this long-standing puzzle would fall into place. Astronomers thus eagerly awaited GCRO's first results for bursts but were astonished to find the same random sky distribution as before. Indeed, the BATSE data appeared to put our Solar System near the center of a spherical distribution of bursting sources.

Another possibility entertained by astronomers was that the bursts originated in the Oort cloud, that vast swarm of comets extending about half a light year from the Sun in all directions. But nothing observed in the Solar System could produce the energies needed to manifest a gamma ray burst. Others envisaged that these bursts might occur in the aftermath of a collision between one or more comets which had collided with neutron stars that had strayed too close. Alternatively, if the bursts represented a more distant population of neutron stars, they might be expected to be concentrated across hundreds of thousands of light years throughout the Galactic halo. The halo would have to be that large, or our vantage some 25,000 light years from the center of the Milky Way would cause the bursts to cluster in that direction. But the results showed that they did not. Thus astronomers were faced with yet another puzzle; how did the neutron stars wind up in these locations?

In the end, astronomers were faced with an altogether different possibility; gamma ray bursts could not have originated with neutron stars at all but instead were derived from deep space, out among the galaxies of the early universe and in a process of extraordinary violence. The BATSE results prompted a major revision of our understanding of the origin and nature of gamma rays bursts.

One of the key mission goals of GCRO's maiden year in Earth orbit was to map the entire sky's gamma-ray emissions. And about halfway through this survey, the EGRET team published the detection of some 16 galaxies exhibiting gamma ray emissions with energies greater than 100Mev. These were no ordinary galaxies however, like our own Milky Way for example, which radiates a fairly steady flux of energy, mainly at visible wavelengths. Gamma ray bursts seemed to be strongly correlated with the so-called active galaxies, with enormous amounts of energy observed emerging from them across the electro-magnetic spectrum from radio waves to gamma-rays. What's more, these active galaxies were shown to be highly variable in terms of their energy output. One of the first objects shown by EGRET to be gamma-ray galaxy was 3C 279, in Virgo, a variable quasar located about 7 billion light years from Earth. Seen in gamma radiation alone, 3C 279 appeared to be radiating some 20,000 times more energy than our Galaxy emits over the entire EM spectrum. Intriguingly, 3C 279 was not detected by gamma-ray satellites in orbit during the 1970s and 1980s, even though they were sensitive enough to pick up such a beacon. The only explanation was that these gamma ray bursts had been activated only recently and might even enter a quiescent phase once again.

The discovery of 3C 279 and other active galaxies not only showed that they were more luminous than normal galaxies at all wavelengths, but that they were observed to radiate the majority of their energy as gamma rays. Of the 16 galaxies discovered by GCRO in its first year of its mission, the most distant objects were estimated to about 11 billion light years away, and the most energetic outshone the Milky Way by five orders of magnitude! As a result, it appeared likely that many more galaxies like those in the very distant universe would be unveiled by the time EGRET completed its all-sky survey.

As one might expect, no shortage of theories were proposed to explain the extraordinary luminosity and great variability of these gamma ray sources. One of the more likely explanations implicated supermassive black holes—with 10 billion or more solar masses—that could accrete matter and heat it up to a trillion degrees, releasing a torrent of gamma rays in the process. Bizarrely, in order to explain the observed variability over a period of several days or so would require the conversion of one solar mass into pure energy, in accordance with Einstein's famous equation linking mass and energy, $E = Mc^2$.

AS GCRO's primary mission came to an end, more than 170 unidentified gamma ray sources were mapped across the sky. GCRO also conducted the first gamma ray survey of the galactic core, and as a result discovered a possible antimatter cloud in orbit right above the center. GCRO recorded more than 2700 gamma ray bursts, at an average rate of discovery of about one per day. Compton was the first to uncover so-called soft gamma ray repeaters—those of unpredictable, short duration that had previously been impossible to detect. Indeed, the wealth of data collated by GCRO forced astronomers to sub-classify the various kinds of gamma ray sources. Even on Earth, gamma ray sources were also recorded during the most violent thunderstorms.

In the distant universe, GCRO studied extremely distant quasars, some 11 in all, finding that that they radiate enormous quantities of gamma rays—sometimes up to four or five orders of magnitude brighter than all other wavebands combined! (Fig. 5.3).

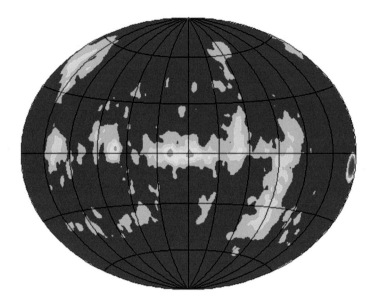

FIG. 5.3 The all sky survey map produced by the COMPTEL instrument onboard GCRO. The prominent *green areas* running through the center reveal the plane of the Milky Way. *Red spots* indicate areas of intense gamma ray emission (Image credit: NASA)

It is all the more remarkable that during a mission lasting a mere nine years, our understanding of the gamma ray universe had been totally revolutionized. But all good things come to an end, and so it was with Compton. A gyroscope aboard GCRO failed in 2000, essential for steering and moving the spacecraft. While all the observational instruments were still running perfectly, with limited maneuverability still possible, the malfunctioned gyroscope created a new danger. Compton could have fallen at any time into the atmosphere and crashed into human settlements, which is not exactly most people's idea of a safe end to a mission. So, on June 4, 2000, NASA deliberately deorbited the telescope, allowing most of it to burn up during re-entry and what was left of it to crash safely into the Pacific Ocean.

The CGRO provided large amounts of data which are still being used to improve our understanding of the high-energy universe, and perhaps the most important of which were the mysterious gamma ray bursts; short lived by highly energetic emissions of gamma rays lasting anywhere from a tiny fraction of a second to over 30 seconds in duration (Fig. 5.4).

2704 BATSE Gamma-Ray Bursts

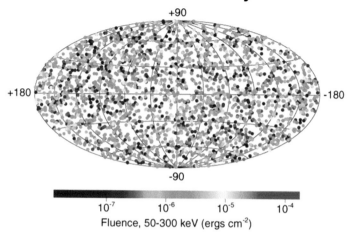

Fluence, 50-300 keV (ergs cm^{-2})

FIG. 5.4 The position of GRBs as mapped by the BATSE telescope on board the GCRO. Note how the bursts are spread uniformly across the sky and are not concentrated along the mid plane of our Galaxy as was previously thought (Image credit: NASA)

After GCRO was deactivated, NASA launched its High Energy Transient Explorer-2 (HETE-2) in October 2000 to continue the important work begun by its highly successful predecessor. It had an improved pointing accuracy of 10 arc minutes at gamma ray wavelengths and a mere 10 arc seconds in X-ray wavebands. What's more, advances in technology enabled HETE-2 to instantly relay the positions of gamma ray bursts to other space telescopes and ground based observatories, enabling rapid searches for counterparts to be carried out. Since its launch, HETE-2 more than doubled the tally of known bursts and played seminal role in enhancing our understanding of these violent explosions (Fig. 5.5).

As we have seen, many theories were forwarded to explain these bursts, most of which implicated nearby sources within the Milky Way Galaxy. BATSE provided crucial data that showed the distribution of GRBs is isotropic, that is, not biased towards any particular direction in space. Because of the flattened shape of the Milky Way Galaxy, if the sources were from within our own galaxy they would be strongly concentrated in or near the galactic plane. The absence of any such pattern in the case of GRBs provided strong evidence that gamma-ray bursts must come from well beyond the Milky Way.

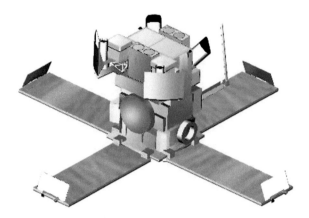

Fig. 5.5 The HETE-2 spacecraft being inspected by engineers at MIT (Image credit: NASA)

What was the cause of these intense bursts of gamma radiation? At first it was thought that they were simply extreme cases of more familiar phenomena like X-ray bursters (described in a later chapter), where matter accreted from a close binary companion was subjected to more violent nuclear burning accompanied by correspondingly higher energy radiation bursts, but in recent years this appears to be not the case.

The smooth distribution (isotropy) obtained from the GCRO convinced astronomers that the bursts could not originate within our own galaxy, as had been widely assumed, but instead must be derived from far beyond the confines of the Milky Way. In other words GRBs must be located at cosmological distances. At the time GCRO was performing its grand survey of GRBs, the distances to these sources could not be determined as there was no standard candle that could be used to estimate their distances. In order to nail their distances accurately, astronomers would have to link GRBs to more familiar objects, like galaxies. But this was easier said than done. In particular, GRBs would have to be monitored at other wavebands, particularly at X-ray and visible wavelengths. The first direct measurement of the distance to a GRB was made in May of 1997. Dubbed GRB 970508, the source was first recorded by the Italian BeppoSAX satellite which detected both the gamma ray and X-ray waves. Using two wavebands rather than one enabled BeppoSAX to pinpoint its location to within a few minutes of arc.

Most models of GRBs invoke a narrow jet of superhot gas radiating furiously in the gamma ray spectrum. The complex burst structure and afterglows are produced as the fireball expands, then cools and interacts with the interstellar medium. If the gamma ray burst is emitted as a jet, then its total energy may be reduced. One good analogy is a laser beam. The total power output from such a laser may be very small, but if this power is concentrated into a beam of very small cross-sectional area, its intensity can be enormous.

Based on these considerations, two models have been proposed to explain GRBs. In one scenario, two neutron stars merge, releasing enormous amounts of energy in a very short time. In a second model, a giant star begins its core collapse just like an ordinary supernova. During this process, the supernova stalls and forms a black hole. Matter that continues to fall into the black hole makes a new accretion disk, which then creates a collimated beam in what astrophysicists now refer to as a hypernova.

Which model prevails today? At the time of writing, this author has learned that the hypernova model seems to be more popularly accepted. In particular, the hypernova model predicts rather long lived bursts, yet some GRBs exhibit some short bursts which seem better explained by the merger of two neutron stars. Thus, it remains possible that there are in fact two sources for these gargantuan explosions from the distant universe.

The most recent research suggests that the frequency of GRBs has decreased over time. This means that in the distant past these violent explosions were significantly more frequent than in the present epoch. Scientists involved in searching for intelligent extra-terrestrial intelligent life beyond the Earth have become duly concerned about the implications of these findings. If a habitable planet finds itself in the beam of a GRB, the radiation is more than capable of sterilizing its surface, bringing an end to any putative life forms that may inhabit that world. Indeed, researchers have pointed out that there is a 95 % chance that the Earth experienced a GRB sterilizing event at some time in its history, leading to one or more mass-extinction events. Perhaps GRBs provide a good answer to the Great Silence (discussed later) currently observed in our Galaxy. In other words, GRBs provide an explanation of why our civilization may be alone in all the cosmos. Such a thought is at once terrifying as it is thought provoking.

The discovery of GRBs impelled astronomers to design and launch new gamma ray telescopes to probe ever deeper into the gamma ray sky. In 2002, ESA launched their INTErnational Gamma-Ray Astrophysics Laboratory (INTEGRAL) in collaboration with the Russian Space Agency and NASA. INTEGRAL's highly eccentric orbit had a period of 72 hours, with its perigee within the magnetosphere radiation belt close to the Earth at 10,000 km. However, most of each orbit was spent outside this region, where scientific observations may take place. It reaches a furthest distance from Earth (apogee) of 153,000 km. The apogee was placed in the northern hemisphere, to reduce time spent in eclipses, and maximize contact time over the ground stations in the northern hemisphere.

The instruments carried by INTEGRAL included IBIS (Imager on-Board the INTEGRAL Satellite) which observed from 15 keV (hard X-rays) to 10 MeV (gamma rays). The angular resolution of IBIS was 12 arcminutes, enabling a bright source to be located to better than 1 arc minute.

The primary spectrometer aboard INTEGRAL was called SPI, the SPectrometer for INTEGRAL. It was conceived and assembled by the French Space Agency (CNES) and observes radiation between 20 keV and 8 MeV.

INTEGRAL really exceeded its 2.2-year planned lifetime. Barring mechanical failures, it is anticipated to continue to function until December 2016. Its orbit was adjusted in early 2015 to cause a safe (southern) re-entry planned for February 2029. Indeed, there should be sufficient fuel left for science operations past 2020 if ESA judges that the scientific return of the missions continues justifying its operating costs (Fig. 5.6).

INTEGRAL was the first space observatory that could simultaneously observe objects in three regions of the electromagnetic spectrum; gamma rays, X-rays and visible light. Its principal targets are violent GRBs, supernova explosions, and regions in the universe thought to contain black holes.

The orbiting observatory has mapped the galactic plane in gamma-rays, resolved diffuse emanating from gamma-ray emission from the galactic center and provided supporting evidence for tori inside active galactic nuclei. To date, it has detected well over 100 of the brightest supermassive black holes in other galaxies.

FIG. 5.6 Artist's impression of the INTEGRAL satellite in Earth orbit (Image credit: ESA)

The Swift Gamma Ray Burst Mission

Interest in GRBs among members of the astronomical community increased so much that researchers wished for a dedicated space observatory monitoring these powerful explosions from the across the universe. A team of astronomers and engineers from the USA, Italy and the U.K were assembled to lay down plans for a new robotic mission that could not only detect GRBs but also follow their evolution as they cooled off. This would require instruments not only sensitive to gamma rays but others that could monitor these bursts at other regions of the electromagnetic spectrum.

Built by the American aerospace company Spectrum Astro, the Swift Gamma Ray Burst spacecraft was launched on November 20, 2004 on a Delta II rocket launcher from Cape Canaveral. Swift carried three instruments.

The Burst Alert telescope (BAT): This instrument detects GRB events and computes its coordinates in the sky. It locates the position of each event with an accuracy of 1 to 4 arc-minutes within 15 seconds. This crude position is immediately relayed to the ground, and some wide-field, rapid-slew ground-based telescopes can catch the GRB with this information. The BAT uses a coded-aperture mask of 52,000 randomly placed 5 mm lead tiles and covers an energy range between 15–150 keV. BAT was designed to monitor large swathes of the sky.

The X-Ray telescope (XRT): This takes images and performs spectral analysis of the GRB afterglow and provides a more precise location of GRB candidates, with a typical error radius of approximately 2 arc seconds. The XRT is also used to perform long-term monitoring of GRB afterglow light-curves for days to weeks after the event, depending on the brightness of the afterglow. The XRT uses a Wolter (grazing incidence technology described in Chap. 1) Type I X-ray telescope with 12 nested mirrors, focused onto a single MOS charge-coupled device (CCD). On-board software allows for fully automated observations, with the instrument selecting an appropriate observing mode for each object, based on its measured count rate. The telescope has an energy range of 0.2–10 keV.

The Ultraviolet/Optical Telescope (UVOT): This detects an optical afterglow. The UVOT provides a sub-arcsecond positioning and conducts optical and ultra-violet photometry through lenticular filters and low resolution spectra (170–650 nm) through the use of its optical and UV grating prisms (grisms). UVOT provided long-term follow-ups of GRB afterglow light curves.

The principal investigator assigned to Swift is Neil Gehrels (NASA Goddard Space Flight Center) and operated from a ground station at the University of Pennsylvania State University (Fig. 5.7).

After suffering a few glitches in the weeks and months after its launch, Swift began to send back prodigious amounts of new data on GRBs and other astrophysical phenomena. In May 9, 2005, Swift detected GRB 050509B, a burst of gamma rays that lasted one-twentieth of a second. The detection marked the first time that the accurate location of a short-duration gamma-ray burst had been identified and the first detection of X-ray afterglow in an individual short burst.

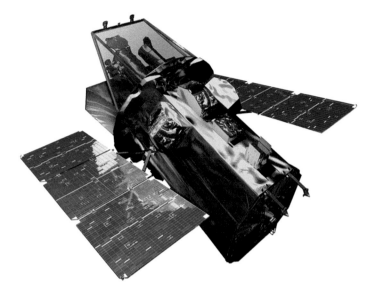

Fɪɢ. **5.7** The Swift Robotic gamma ray observatory (Image credit: NASA)

This was followed in September 4, 2005 with the discovery of GRB 050904 with a redshift value of 6.29 and of unusually long duration (200 seconds where most of the detected bursts persist for only 10 seconds). It was also found to be the most distant yet detected, at approximately 12.6 billion light-years. On February 18, 2006, Swift detected GRB 060218, another unusually long (about 2000 seconds) and nearby (about 440 million light-years) burst. The source was unusually dim despite its close distance, and may be an indication of an imminent supernova.

On June 14, 2006, Swift detected GRB 060614, a burst of gamma rays that lasted 102 seconds in a distant galaxy (about 1.6 billion light-years). Curiously, no supernova was seen following this event leading some to speculate that it represented a new class of object. Others suggested that these events could represent massive star deaths, but ones which produced too little radioactive Nickel 56 to power a supernova explosion.

On January 9, 2008 Swift was observing a supernova known as NGC 2770 when it detected an X-ray burst coming from the same galaxy. The source of this burst was found to be the beginning of another supernova, later called SN 2008D. This result was notable in that never before had a supernova been seen at such an early stage in its evolution. Following this serendipitous discovery, astronomers were able to study in detail this Type Ibc supernova with an extensive range of telescopes including the Hubble Space Telescope, the Chandra X-ray Observatory, the Very Large Array in New Mexico, the Gemini North telescope in Hawaii, Gemini South in Chile, the Keck I telescope in Hawaii, the 1.3 m PAIRITEL telescope at Mt Hopkins, the 200-inch and 60-inch telescopes at the Palomar Observatory in California, and the 3.5-meter telescope at the Apache Point Observatory in New Mexico.

In more recent years, Swift has gone on to discover over a thousand other GRBs. In addition, its sensitive instrumentation was able to provide valuable insights into other types of high-energy astrophysical phenomena. For example, on April 23, 2014 Swift detected the strongest, hottest, and longest-lasting sequence of stellar flares ever seen from a nearby red dwarf star. The initial blast from this record-setting series of explosions was as much as 10,000 times more powerful than the largest solar flare ever recorded. The spacecraft continues to relay new discoveries to ground based researchers on a regular basis.

The Fermi Gamma Ray Telescope

The latest and greatest gamma ray telescope ever to be launched is the Fermi satellite, launched on June 11, 2008. The Fermi mission represents a collaborative venture between NASA, the United States Department of Energy, and government agencies in France, Germany, Italy, Japan, and Sweden. Its main instrument is the Large Area Telescope (LAT), with which astronomers mostly intend to perform an all-sky survey studying astrophysical and cosmological phenomena such as active galactic nuclei, pulsars, other high-energy sources and dark matter. Another instrument aboard Fermi, the Gamma-ray Burst Monitor (GBM; formerly GLAST Burst Monitor), is employed to study gamma-ray bursts (Fig. 5.8).

The key scientific objectives of the Fermi mission are to understand the mechanisms of particle acceleration in active galactic nuclei (AGN), pulsars, and supernova remnants (SNRs). In addition, Fermi is anticipated to produce the highest resolution images of the gamma-ray sky, helping elucidate unidentified sources and

FIG. 5.8 The Fermi gamma ray telescope awaiting launch at Cape Canaveral (Image credit: NASA)

diffuse emission. In addition, it is hoped that Fermi will help determine the high-energy behavior of gamma-ray bursts and transients.

In addition to these mission goals, Fermi will help throw new light on the nature of dark matter by looking for so-called gamma ray excess from the center of the Milky Way, as well as from galaxies in the early universe. Other researches have commissioned time with Fermi to search for evaporating primordial micro black holes (MBH) from their presumed gamma burst signatures.

Perhaps the most amazing discovery made by Fermi to date is a gigantic, mysterious structure in our galaxy. This never-before-seen feature looks like a pair of bubbles extending above and below our Galactic center (Figs. 5.9 and 5.10).

At the time of writing, Fermi continues to produce fascinating new data on many types of high-energy astrophysical phenomena and is anticipated to continue working well into 2018 and beyond.

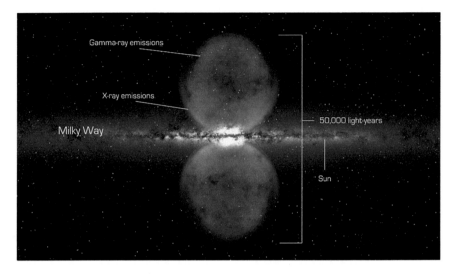

FIG. 5.9 Schematic showing the giant gamma ray lobes discovered extending above and below the plane of our galaxy (Image credit: NASA)

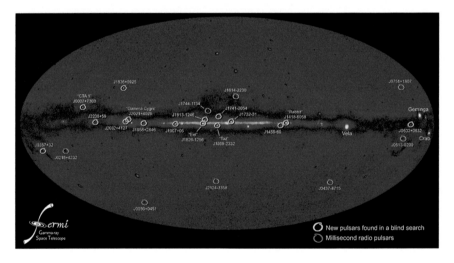

FIG. 5.10 Fermi map of Gamma ray pulsars discovered by Fermi in this all sky map (Image credit: NASA)

In the next chapter, we shall explore the universe as revealed by ultraviolet eyes, and we'll discover how space-based observatories that can see at these wavebands have added a new layer of complexity onto the already fascinating cosmic arena.

6. The Universe Through Ultraviolet Eyes

Ultraviolet (UV) light (or radiation) has shorter wavelengths than visible light. Though UV waves are well-nigh invisible to the human eye, some insects, birds and fish can see them! Scientists have divided the ultraviolet part of the spectrum into three regions: the near ultraviolet, the far ultraviolet, and the extreme ultraviolet. The three regions are distinguished by how energetic the ultraviolet radiation is. The longer the wavelength, the less energetic the radiation. The near ultraviolet, abbreviated NUV, is the light closest to optical or visible light, and covers a wavelength range between 300 and 400 nm. The far ultraviolet, abbreviated FUV, lies between the near and extreme ultraviolet regions (122–200 nm). It is the least explored of the three regions. The extreme ultraviolet, abbreviated EUV, is the ultraviolet light closest to X-rays, and is the most energetic of the three types (10–121 nm).

Our Sun emits light at all the different wavelengths in electromagnetic spectrum. Most of this radiation is blocked by the atmosphere, but some ultraviolet waves do manage to get through, causing sunburn. Ultraviolet radiation has wavelengths of about 400 nanometers (nm) on the visible-light side and about 10 nm on the X-ray side. Earth's stratospheric ozone layer blocks all wavelengths shorter than 300 nm from reaching ground-based telescopes. As this ozone layer lies at an altitude of 20–40 km (12–25 miles), astronomers have to resort to rockets and satellites to make observations below this wavelength range. Some days, more ultraviolet waves get through our atmosphere. As a result, scientists have formulated a UV index to help people protect themselves from these harmful ultraviolet waves.

We can study stars and galaxies by observing the UV light they emit. The Far UV Camera/Spectrograph deployed and left on the Moon by the crew of Apollo 16 took this picture (Fig. 6.1).

© Springer International Publishing Switzerland 2017
N. English, *Space Telescopes*, Astronomers' Universe,
DOI 10.1007/978-3-319-27814-8_6

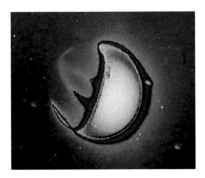

Fɪɢ. 6.1 The Earth as imaged from the Moon using the Apollo 16 UV spectrograph (Image credit: NASA)

As one can see, the part of the Earth facing the Sun reflects much UV light. Even more interesting is the side facing away from the Sun. Here, bands of UV emission are also apparent. These bands are the result of aurora caused by charged particles given off by the Sun. They spiral towards the Earth along Earth's magnetic field lines.

Many scientists are interested in studying the invisible universe of ultraviolet light, since the hottest and the most active objects in the cosmos give off large amounts of ultraviolet energy. The difference in how the galaxies appear is due to which type of stars shine brightest in the optical and ultraviolet wavelengths. Images of galaxies show mainly clouds of gas containing newly formed stars many times more massive than the sun, which glow strongly in the ultraviolet. In contrast, visible light pictures of galaxies show mostly the yellow and red light of older stars. By comparing these types of data, astronomers can learn about the structure and evolution of galaxies.

Ultraviolet line spectrum measurements are used to discern the chemical composition, densities, and temperatures of the interstellar medium, and the temperature and composition of hot young stars. UV observations can also provide essential information about the evolution of galaxies.

The ultraviolet universe looks quite different from the familiar stars and galaxies seen in visible light. Most stars are actually relatively cool objects emitting much of their electromagnetic radiation in the visible or near-infrared part of the spectrum.

Ultraviolet radiation is the signature of hotter objects, typically in the early and late stages of their evolution. If we could see the sky in ultraviolet light, most stars would fade in prominence. We would see some very young massive stars and some very old stars and galaxies, growing hotter and producing higher-energy radiation near their birth or death. Clouds of gas and dust would block our vision in many directions along the Milky Way.

Like the other forms of electromagnetic radiation, UV astronomy had its beginnings early on in the space programs developed by the USA, Soviet Union and the nations of Europe. One of the first satellites launched to explore the UV universe was Thor-Delta 1A (TD-1A) a European astrophysics research satellite first launched into space in 1972. Operated by the European Space Research Organisation (ESRO), TD-1A made astronomical surveys primarily in the ultraviolet, but it also carried X-ray and gamma ray detectors. The Stellar UV Radiation Experiment, operated by University College London and the University of Liège, consisted of a 1.4-metre ultraviolet telescope, with a sensitive spectrometer attached. It was used to study extinction and to produce a star catalogue using ultraviolet observations over the range 135 to 255 nm.

TD-1A carried a UV Stellar Spectrometer , designed and operated by the Astronomical Institute of Utrecht University, consisted of a diffraction grating spectrometer attached to a 26 centimeter aperture Cassegrain reflector telescope and was employed to study UV radiation at 216, 255 and 286 nm. TD-1A operated for a little over two years, conducting over two complete sky surveys in that time, and mapping some 95 % of the entire sky. TD-1A was able to scan a thin band of the sky on each orbit, by keeping its orientation fixed with respect to the Sun. In addition by trailing Earth's orbital motion around the Sun, it was able to scan the entire sky over a period of six months.

The efficacy of TD-1A was ultimately compromised by a reliability issue with an on-board tape recorder, which was used to store experimental data for later transmission back to ground stations on Earth. Luckily, these malfunctions were intermittent, allowing the spacecraft to complete its mission. Further problems with an encoder led to the postponement of the Spectrometry of Celestial X-rays experiment during the first survey, but this was rectified in enough time for its operation during the second survey, and so the mission was still completed successfully. In May 1974, TD-1A gave up the ghost when its attitude control system ran out of propellant, leaving it unable to maintain its orientation. The spacecraft decayed from its orbit and re-entered the atmosphere in a blaze of glory on January 9, 1980.

The UK's European partner, the Netherlands, soon developed their own astronomical satellite designed to explore the ultraviolet universe. Known as Astronomische Nederlandse Satelliet (ANS), it was a combined X-ray and ultraviolet telescope. Launched into Earth orbit on 30 August 1974 at 14:07:39 UTC astride a Scout rocket from Vandenberg Air Force Base, United States, the mission ran for 20 months until June 1976, and was jointly funded by the Netherlands Institute for Space Research (NIVR) and NASA (Fig. 6.2).

Remarkably, ANS completed all of its UV astronomy with a rather small 22 cm Cassegrain telescope. The wavelengths observed fell into the range between 150 and 330 nm, with the

Fig. 6.2 Artist's impression of the ANS satellite in Earth orbit (Image credit: Wikimedia Commons:commons.wikimedia.org/wiki/File:ANS_backup_flightarticle.jpg)

detector dividing up into five channels with peak wavelengths of 155, 180, 220, 250 and 330 nm. At these frequencies it took over 18,000 measurements of around 400 objects.

Not to be outdone, the USA designed a suite of small UV telescopes for flight on board its Orbiting Astronomical Observatory (OAO) program, which were launched into space over a nine year period, between 1972 and 1981. These missions served more as 'proof of concept' for later, more ambitious space telescopes like the Hubble Space Telescope.

The International Ultra-Violet Explorer

In the early 1970s, astronomers from NASA, the European Space Agency (ESA) and Britain's Science and Engineering research Council (SERC) developed plans for a small ultraviolet satellite—the International Ultraviolet Explorer (IUE)—that would act spectroscopically, collecting data over the wavelength range of approximately 100 nm to 300 nm. The engineers assigned to the project designed the craft to have a three year lifespan, but as luck would have it, the IUE would survive to five times that age and beyond, to become one of the most productive orbiting observatories ever conceived.

The IUE consisted of a 45 cm Cassegrain telescope armed with two spectrographs, and TV cameras to record spectra. The UV spectral data, which had a resolving power of 2 arc seconds, were transmitted to one of two ground stations. IUE was controlled by NASA's Goddard Space Flight Center in Maryland for 16 hours out of every day. For the remaining 8 hours, IUE was monitored from Villafranca, Madrid, Spain.

Launched on January 26, 1978, IUE was placed in a geosynchronous orbit (with a 24 hour period) which maintained its position over the Atlantic at all times. This stable, geosynchronous orbit enabled IUE to be operated in a unique way for spaceborne telescopes, allowing astronomers to conduct observations in much the same way as they used ground-based telescopes, that is, in real-time. For the next 15 years, over 2000 scientists of many nationalities obtained valuable telescope time using IUE's data.

The IUE's exceptionally stable orbit enabled astronomers to make a rapid response to sudden or unpredictable targets of opportunity, such as comets, novae and supernova explosions. For the first time in the history of space science, it was possible to make schedule changes within hours rather than days. What is more, IUE was able to observe serendipitous targets simultaneously with other space observatories surveying the sky at different bands of the electromagnetic spectrum, such as Einstein, ROSAT and Ginga, as well as with large ground-based optical and radio telescopes. This kind of collaboration was to become the rule rather than the exception for many future space-based astronomy programs, enabling astronomers to glean far more information about a celestial target than could previously be achieved (Fig. 6.3).

The IUE was put to work studying all the planets of our solar system with the exception of Mercury. This was because the telescope was prevented from observing any object within 45° of the Sun, with Mercury's greatest distance from the Sun being about 28°. When IUE was turned on Venus it showed that the levels of the oxides of sulfur in its atmosphere fell significantly during the 1980s. The reason for this decline was never fully elucidated, but one the leading hypothesis that volcanic activity on the planet had declined at this time.

When the IUE was directed towards Jupiter, it detected vigorous evidence of auroral displays in its upper atmosphere, the most prominent being near its north and south poles. The same was true of Saturn. In addition, IUE was able to sniff out highly energetic molecular species in the atmospheres of all the major planets greatly increasing our knowledge of the chemistry of the planets that orbit our Sun.

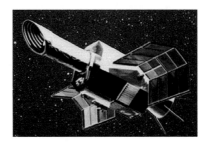

Fɪɢ. 6.3 Artist's depiction of the IUE satellite in orbit (Image credit: ESA)

In July 1978, Rutherford Appleton Laboratory scientist David Stickland had just completed a set of calibration exposures and re-directed the IUE, turning the satellite toward one of the most luminous stars in the galaxy, a G-type supergiant star, HR 8752. This star had been distinguished from others by its extreme rate of mass loss and variability. To his astonishment, Stickland's found that, after a long spell of image processing, HR 8752's spectrum was clearly O or early B rather than G! Stickland expressed great surprise at this result but it also raised alarm bells in his head. What if, in fact, he had been observing the wrong star? If so, the spacecraft's attitude might have been off, with the result that the next team signed up to use it would have to correct for its pointing errors.

Fortunately for Stickland, the conundrum had nothing to do with inaccurate positioning. The British astronomer had discovered a hot companion that simply outshone the cool G star at ultraviolet wavelengths. Its faintness at visible wavelengths and close proximity to the primary G-star hid the companion in its glare. This is just one of many amazing stories that was unveiled as the IUE carried out its space mission over a decade and a half, and like all good science, raised more questions than it answered.

One of the most important targets for the IUE mission was to monitor Supernova 1987A in the Large Magellanic Cloud (LMC). First detected on February 2, 1987 by Ian Shelton and Oscar Duhalde at the University of Toronto site in Chile, the astronomical duo were forced to drive down from the mountaintop to a telegraph office several hours away in order to send the report to the late Dr. Brian Marsden, a world expert in celestial mechanics, based at the International Astronomical Union's Central Bureau for Astronomical Telegrams. When IUE project scientist Yoji Kondo (NASA-Goddard Space Flight center) heard the news from a colleague of the discoverers, he phoned the principal investigator for NASA's target of opportunity program for supernovae, Robert Kirshner (Harvard Smithsonian Center for Astrophysics). At that very moment, Kirshner was reaching for a phone to call Kondo.

Kirshner had heard about the supernova event from J. Craig Wheeler (University of Texas). A few years earlier, Wheeler and a few happy-go-lucky friends played a trick on Kirshner by sending a spoof telegram announcing the discovery of a bright supernova in the Andromeda galaxy. So when Wheeler called, Kirshner was understandably skeptical. But when he looked in Marsden's office down the hall, the news of the supernova in the LMC was just arriving from Chile. So both Kondo and Kirshner received the news through informal channels before the discoverers' telegram arrived! NASA and ESA quickly organized an around-the-clock observing program lasting four days, and which actually turned out to be of pivotal importance because of the rapid fading of the supernova at ultraviolet wavelengths.

IUE was the first professional telescope anywhere to observe SN 1987A after its discovery, since no other telescope was situated in the right location on Earth for several hours. The event was so luminous in the far-ultraviolet that the first spectrum recorded by IUE was badly over-exposed. The second one however, was much better. In fact, the supernova was intense enough to allow several high-resolution spectra to be obtained, thereby producing the first ultraviolet atlas of interstellar matter in the large Magellanic Cloud, the Milky Way, and its extensive halo.

Although the far-ultraviolet brightness of SN 1987A subsided to a small fraction of 1 % of its peak value within four days of its discovery, dipping below IUE's sensitivity threshold, the spectra of two other stars could still be discerned in the data. Careful analysis of a pre-explosion photograph showed that there had indeed been three stars close together in the field. Astronomers then compared this photograph to the separation and orientation of the two remaining stars determined by the IUE spectra. In so doing, they were able to confirm that a blue supergiant was now absent, the very star that went supernova. This was the first time that a super-nova precursor star had been unambiguously identified.

IUE's discovery of the companion to the star HR 8752 was not its only such find. Of particular importance was the detection of hot companions to several Cepheid variables, which play a crucial role in determining the size of the universe. The companions provided more accurate masses and luminosities for these valuable objects.

Similarly, observations made by IUE confirmed that so-called barium stars have white dwarf companions. Barium stars are believed to be the result of mass transfer in a binary star system. The mass transfer occurs while the larger star is on the main sequence. Its companion, the donor star, is usually a more highly evolved carbon-rich red star. These nuclear fusion products are mixed by convection and move upward to its surface. During the next phase, some of that matter pollutes the surface layers of the main-sequence star, as the donor star loses mass at the end of its life, when it continues to evolve into a white dwarf. These systems are being observed at an indeterminate amount of time after the mass transfer event, when the donor star has long been a white dwarf, and the 'polluted' stellar recipient has evolved to become a red giant. During its evolution, the barium star will at times be larger and cooler than the limits of the spectral types G or K. When this happens, the star is usually of spectral type M and may reveal the spectral signatures of element zirconium, zirconium oxide and other simple molecule bands.

Historically though, barium stars posed somewhat of a puzzle, because in standard stellar evolution theory, G and K giants are not far enough along in their evolution to have synthesized carbon and a number of s-process elements, a process that produces heavy atomic nuclei by capturing slow neutron in the aftermath of a supernova explosion. The discovery of the stars' binary nature resolved the conundrum, putting the source of their spectral peculiarities into a companion star which should not have produced such material. This mass transfer mechanism is rather short-lived though. The mass transfer hypothesis also predicted that there should be main-sequence stars which present the spectral signs of barium. At least one such star, HR 107, is known to display the element in the far ultraviolet. In this way the IUE provided brand new insights into stellar evolution, greatly enriching our knowledge of previously obscure phenomena.

The IUE proved indispensable in finding many invisible stellar companions. Another example of this comes from Zeta Aurigae-type binary systems. If conditions are favorable, a small, hot stellar component is observed to pass behind an elongated companion and, as a result, becomes progressively dimmed by the layers of gas in the latter's atmosphere. IUE was able to carefully follow these eclipse events—and thus provide valuable data about the internal constitution of their interiors.

IUE imaged Comet Halley from September 1985 through July 1986 as it approached the Sun, when it was at a distance of 2.7 astronomical units (A.U), following it out to 2.8AU on its outward journey away from the Sun back into the depths of interplanetary space. While it was nearest the Sun, IUE identified the spectral signature of hydroxyl ions, which indicated that the comet was losing about one million cubic meters of water each and every day! At this rate of water loss, just a few more perihelic passages would leave Comet Halley quite dehydrated.

Other IUE discoveries were based on previous observations carried out by the satellite. For instance, early in its mission, IUE clearly showed that Seyfert galaxies had highly variable emission lines, the intensities of which appeared to fluctuate in sync with the galaxies' overall brightness changes. This variability allowed astronomers to study the inner regions of active galactic nuclei (AGN), which, as we have seen, were suspected to have central black holes.

A group of 20 scientists convening at an interdisciplinary gathering in 1988 at Segovia, Spain, concluded that IUE was ideally suited to study Active Galactic Nuclei (AGN). They formed an unofficial organization, known as the AGN Watch, to convince the IUE peer-review committees in Europe and the United States to dedicate a significant amount of the satellite's time to this aim. The AGN Watch recommendations were fully endorsed, with both ESA and NASA dedicating the requested observing time— some 700 hours in all—over a period of eight months, to study the Seyfert 1 galaxy NGC 5548 in the constellation of Bootes. The program attracted many professional astronomers, with the final effort involving some 60 scientists from 20 nations. The IUE data revolutionized our understanding of AGNs, so much so that a completely new model had to formulated to explain all of the features of these highly energetic objects.

The IUE was also employed to carry out new observations of the Interstellar Medium (ISM). The ISM contains varying amounts of gas and dust which absorbs light from hot stars in the vicinity. One of the first discoveries of the IUE was that our galaxy, the Milky Way, is embedded in a vast halo of superheated gas, known as a galactic corona. Heated by cosmic rays and supernovae, this gaseous halo extends several thousand light years above and below its mid-planc.

The IUE mission also proved instrumental in determining how the light from distant sources is affected by dust along our line of sight. Almost all astronomical observations are affected by this interstellar extinction, and this requires correcting for in most analyses of astronomical spectra and images. IUE data was used to show that within the galaxy, interstellar extinction can be well described by a few simple equations. The relative variation of extinction with wavelength shows little variation with direction; only the absolute amount of absorption changes. Interstellar absorption in other galaxies can similarly be described by fairly simple laws.

The IUE was planned to have a short lifetime of just a few years. However, it lasted far longer than its design called for. And while occasional hardware failures did jeopardize the mission, engineers devised ingenious procedures to circumvent them. In one situation, the spacecraft was equipped with six gyros to stabilise the spacecraft. These failed one by one between 1979 and 1996 and ultimately left the spacecraft with just a single functional gyro. That said, further brain-storming by mission controllers allowed its pointing accuracy to be maintained with two gyros by using the telescope's solar sensor. What is more, stabilization in three axes proved possible even after the fifth failure, by using the solar sensor, the fine error sensors and the single remaining gyro. But the IUE enjoyed its own share of good luck too, in that the telescope systems remained fully functional throughout the mission.

Budget cutbacks at NASA almost led to the termination of the mission in 1995, but instead the operations responsibilities were re-divided, with ESA taking control for 16 hours a day, and GSFC for the remaining 8. The 16 hours per day allocated to ESA was used for science operations, while the 8 hours given to the GSFC was used only for maintenance. In February 1996, a string of further budget cuts forced ESA to terminate all science projects conducted by the satellite. By September of the same year, operations ceased entirely, and by the end of the same month the remaining hydrazine fuel was jettisoned to the vacuum of space, the batteries were drained and turned off, and at 18:44 UT, the spacecraft's radio transmitter was shut down with the result that all communication with it was lost.

At the time of writing, the IUE continues in its geosynchronous orbit, and will remain there indefinitely as it is far above the upper reaches of the Earth's atmosphere where frictional forces could bring it crashing down. Over time, small anomalies in the

Earth's gravity due to its non-spherical shape meant that the telescope tended to drift west from its original location at approximately 70°W longitude towards approximately 110°W. During the mission, this drift was corrected by occasional rocket firings, but since the mission was terminated the satellite has drifted uncontrolled to the west of its former location.

Owing to its very long duration, the IUE has had a major impact on astronomy. Indeed, by the end of its lifetime, it was considered to be, by some margin, one the most successful and productive space observatory missions of all time, its archives providing data for many professional astronomers and over 250 PhD projects across the world. IUE provided data for over 4000 peer-reviewed papers, including some of the most cited astronomy papers of all time. In comparison, the Hubble Space Telescope has now been in orbit for over two decades and its data has been used in almost 10,000 peer-reviewed publications.

The Extreme Ultraviolet Explorer (EUVE)

Following fast on the heels of the hugely successful IUE mission was another U.S. satellite designed to explore the extreme ultraviolet spectrum (<121 nm). From 1992 to 2001, EUVE surveyed the sky for the first time in the extreme ultraviolet (EUV) region between 4 and 76 nm. The extreme ultraviolet is defined to be between about 10 and 100 nm. It had four telescopes with gold-plated mirrors, the design of which was critically dependent on the transmission properties of the filters used to sample the EUV band passes. The combination of the mirrors and filters was selected to maximize the telescope's sensitivity to detect faint EUV sources. Three of the telescopes had scanners that were pointed in the satellite's spin plane. The fourth telescope, the Deep Survey/Spectrometer Telescope, pointed in a direction opposite the Sun direction and was equipped with three spectrometers.

Launched into space on June 7, 1992, the 3.4 ton EUEVE satellite entered a circular orbit at 550 km altitude inclined some 28 degrees to the Earth's equator. Remarkably, EUVE managed to complete all its survey tasks within a year, before making targeted

FIG. 6.4 Artist's impression of NASA's EUVE satellite in Earth orbit (Image credit: NASA)

spectroscopic observations for scientists from around the world. Astrophysicists at the University of California, Berkeley, together with NASA, operated EUVE around the clock. Beyond the activities of professional researchers, the EUVE had strong ties to the educational and amateur astronomy communities as well. About a third of those responsible for the satellite's operation were students from the University of California. They assisted the guest observer program, kept the computer systems running smoothly and monitored the instruments' status while in orbit. The result was a combination of good education and good economics. In addition, amateur astronomers working through the American Association of Variable Star Observers (AAVSO) provided valuable monitoring of variable sources, as will be described later (Fig. 6.4).

The catalog of objects surveyed by EUVE contained more than 800 sources, about 300 of which were relatively cool, low mass, late-type stars. EUVE was able to observe coronal emission from these stars, which exhibited temperatures in excess of a million Kelvin. Another third of EUVE's sources consisted of white dwarfs with radiation emanating from 25,000 to 100,000 K photospheres, as well as winds of the massive and hot early-type stars, cataclysmic variables, active galactic nuclei (AGNs), as well as such unique objects as the Vela supernova remnant, the Cygnus Loop, and much closer to home; the Jupiter-Io torus.

When space astronomy first became a reality in the late 1950s, astronomers began to consider what wavelengths could be exploited outside the optical and radio bands then available. Unfortunately, the radio data suggested that a great concentration of cool hydrogen permeated the whole Galaxy, making the ISM extremely opaque to EUV radiation. The more neutral hydrogen gas along a line of sight, the more the radiation from a EUV source will be diminished. For instance, if there were one hydrogen atom per cubic centimeter in the ISM, as was thought back then, at 91.2 nm (the wavelength at which hydrogen opacity peaks) we could see less than a light year away—far short of the 4-light year distance to the nearest star, Proxima Centauri. If the average density were reduced to one hydrogen atom per 10 cubic centimeters, one could see ten times farther, yet still only half as far as Proxima Centauri. Such calculations performed 30 years ago did not bode well for EUV astronomy in general, except perhaps at the shortest wavelengths where the opacity was reduced.

The unambiguous results from EUVE showed that extreme ultraviolet sources can be visible out to many hundreds of light years. The underlying reason is that the ISM is not uniformly dense, as was once envisaged. It doesn't permeate space like a fog. Rather it creates a partly cloudy sky. In some areas, the ISM is fairly dense and cold; in others it is less dense and far hotter— instead of having a temperature of a few hundred degrees, some pockets of the ISM can be between 100,000 to a million degrees. Here not only is the overall density lower, but the amount of neutral hydrogen is dramatically less because most of it is ionized and with this the majority of the opacity disappeared.

There are even some patches, or tunnels, between the clouds in which there is so little attenuation that one can see beyond our galaxy, like seeing glimpses of the constellations on a partly cloudy night. And once we have a clear view out of the galaxy in a given direction, we can see to cosmological distances. Indeed EUVE observed 20 AGNs and related objects situated not hundreds, but hundreds of millions of light years away! To complicate this picture further, there appeared to be cool, dense material in the immediate vicinity of the Sun (within a few light years). Our Solar System, together with this cool cloud, apparently resides inside a hot, tenuous bubble in the ISM. Other cool clouds are thought to be scattered throughout the hot region, as suggested by a comparison of the diffuse EUV background with infrared maps showing cool neutral material.

It should not surprise the reader that the main categories of objects visible in the EUV also emit strongly in the neighboring X-ray region of the EM spectrum. For instance, main-sequence stars of spectral types F through M have X-ray emitting coronas just like the Sun, as do all evolved F through early K giants and supergiants. On the other hand, cooler mid-K and M giants are not coronal sources and seem to be largely invisible in the EUV.

Based on their best knowledge of the Sun, astronomers believed stellar coronas consist mainly of hot plasmas—a hot soup of charged particles—with temperatures of between 1 to 10 million degrees restrained by magnetic fields in coronal loops. Plasmas at these temperatures and densities radiate in both UV and soft X-rays, primarily as line emissions. The latter offered many new insights about coronal conditions. Two of the first active stars for which EUVE obtained coronal spectra were the Capella binary (two type-G stars) and the Sun-like star Chi[1] Orionis (these have strong emission from the equivalent of active regions on the Sun.) The brightness of these stars were monitored during the long spectral exposures (which often lasted several days), to see if emission was steady or if flares were emitted. However, neither Capella nor Chi[1] Orionis showed any flaring during the exposures, yet they both had spectral lines indicating temperatures of 10 million degrees or more.

On the Sun, such temperatures are only reached during flaring events. However, the Yohkoh (discussed in a later chapter) spacecraft showed that there was constant, low level, flare-like activity on the Sun, and it could explain the presence of such high-temperature emission lines in the spectra of these two stars. The physicist Jurgen Schmitt (Max Planck Institute), and his colleagues, showed that Chi[1] Orionis were copious emitters of coronal radiation because it was very young, perhaps only 300 million years old (compared to the Sun's 4.5 billion years). It was observed to rotate some three or four times faster than the Sun. Such rapid rotation combined with convection beneath the surface of a star, are believed to give rise to magnetic fields through an internal dynamo process. In general, a more rapid the rotation results in a stronger dynamo effect.

On the other hand, both stars in the Capella binary system have long since left the main sequence. Their close, 104-day orbit tidally locks them together in such a way that they rotate somewhat faster than they would do in isolation. And while their rotation periods are still slower than the Sun's, their radii are much larger, with the result that these giant stars have much higher surface velocities. This pair belongs to the class of binaries known as RS Canum Venaticorum (RS CVn) stars, which overall are the most active of stellar objects and exhibit the most intense EUV coronal emissions.

One of the most energetic RS CVn stars is the binary system known as V711 Tauri. Researchers studying its EUV emission had found intensity variations up to 40%, which correlated with the three day orbital period of the pairing. As with the Sun, activity of these stars is not evenly distributed over their surfaces. Spots and surrounding active regions turn up and then disappear in equatorial and mid-latitude bands and are carried into and out of view by virtue of their rotation. In optical light, active regions are dark but in the ultraviolet an X-ray bands, they are shown to be bright.

The rotational characteristics of V711 Tauri betray the presence of long-lived, compact regions of exceptionally high emission on the surface of one of the two stars. The geometry of this system provides an indirect way of mapping the surface of what would otherwise be a point source. As one might expect, RS CVn stars also produce flares of enormous dimensions—sometimes as much as 1000 and 10,000 times more energetic than those observed on the Sun. In some instances, they have been known to last more than nine hours. The next most active flare stars are those which present with very cool photospheres—the M-class dwarfs like Proxima Centauri. In these cases, flares 100 times as intense as the Sun's are ejected from stars that are only one thousandth as luminous! Several such events on these stars were detected during EUVE's all-sky survey, when stars were periodically scanned during each orbit.

One of these took place on the M dwarf, AU Microscopii, during an instrument-calibration exercise. Since each photon is time tagged, the spectrum was easily separated into 2 phases a quiescence stage followed by a flaring stage, where peak emissions typically lasted for two hours and where the emission took up to 24 hours to dissipate. Both the flare's time profile and spectrum allowed a team of astronomers led by Scott Culley and George Fisher (University of California, Berkeley) to suggest that the event was a scaled up version of a solar coronal mass ejection but the amount of mass and energy involved are about 10,000 times as much as for a similar event on the Sun.

That a little M dwarf may undergo such violent outbursts is of more than passing interest. For instance, a star may be decelerated or spun down by many such outbursts. A second consequence follows from the sheer numbers of galactic M dwarfs in existence. If these mass ejections occur on a significant fraction of these stars, the ISM may have originated mostly from such outbursts, surpassing planetary nebulae, supernovae, and other stellar winds as the leading contributor.

As we have seen, the higher the temperature of a black body, the shorter its peak wavelength of emission. This rule, known as Wien's law (discussed in Chap. 1) predicts that white dwarfs with surface temperatures between 25,000 and 100,000 K should emit the bulk of their radiation in the EUV. Indeed, these stars are the second most abundant type of EUV object. Astronomers have discovered a puzzling haze of heavy elements in white dwarf spectra. Theories predict that white dwarfs, having come to the end of their thermonuclear evolution, would have extremely stable, stratified atmospheres (unlike the turbulent, convective solar atmosphere) caused by intense gravitational fields, which help fractionate the elements. Heavy elements should settle far below the observable surface, but if the white dwarf is hot enough, radiation pressure pushes certain elements back to the surface. Signatures of this effect are most apparent in the EUV, where they found spectra differing by as much as a factor of 10 from predictions made just a few years previous prior to these observations.

As interesting as white dwarfs are in their own right, dramatic situations can arise when one of them has a close companion. In two examples discovered by Stephane Vennes (University of California, Berkeley) and John Thorstensen (Dartmouth College), one side of a cool, red dwarf is greatly heated by its proximity to its white dwarf companion. Bizarrely, the result is that the heated side of the red dwarf emits a different spectrum than its cool side does! In one such binary system, the red dwarf is just one tenth as far away from the white dwarf as Mercury is from the Sun, and the two stars orbit each other in only 4.23 days.

When a white dwarf and a companion are in a yet tighter orbit, with a period of just a few hours instead of days, even more intriguing interactions occur. Material is pulled off the cool star and accreted onto the white dwarf, creating a so-called cataclysmic variable (CV). The accretion is important in such radically diverse objects as protostars, AGNs, and black holes, and the knowledge gleaned from EUV observations of CVs has greatly illuminated other areas of astrophysics.

For example, EUVE has detected time-variable radiation from a CV known as RE 1938-461. This star possesses a powerful magnetic field estimated at more than 50 million Gauss (in comparison, sunspots have field strengths of about 1000 Gauss, and the Earth, a mere half Gauss) which prevents an accretion disk from forming. The companion's material flows directly onto the CV at the poles, emitting EUV and other high-energy radiation as it does so. Variations in visible light attest to significant changes in the accretion rate, which is now observable in the EUV as well.

In another instance, amateur astronomers played a pivotal role in capturing an outburst on the dwarf nova, SS Cygni, a cool type-K dwarf closely orbiting a hot white dwarf. Many amateurs observed a sudden increase in optical brightness and reported it to the AAVSO, which in turn, quickly relayed the information to NASA. Within 24 hours EUVE was observing this still-rising outburst that lasted for five days. Thanks to careful coordination, astronomers were able to obtain almost simultaneous observations with the IUE.

Much of the EUV radiation from CVs is thought to emanate from the hot spots where material is being accreted. The size of these spots cannot be seen directly with any telescope, but the geometry of binary star eclipses enables astronomers to deduce that information. Paula Szkody (University of Washington) and her colleagues observed the eclipsing system VV Puppis, in which a late-type main sequence star and a white dwarf orbit each other in only 100 minutes. As with RE 1938-461, there is no accretion disk; material funneled directly onto the magnetic poles of the white dwarf. From the spectrum and the eclipses, the astronomers inferred a temperature of 300,000 K and a diameter of only 100 kilometers for the area onto which the material flows.

In another study, a three-day EUVE observation of the UX Fornacis system by John Warren (University of California, Berkeley) and his colleagues appeared to show variations in the stream itself on its way to becoming a white dwarf. The period of this binary is only 126.5 minutes, and EUV observations of 36 orbits chronicled bouts of emission, appearing and disappearing as the hot spot rotated into and out of view, respectively, just like the situation with VV Puppis. But there were some important differences as well—the variations also included dips due to obscuration of the hot spot by the in-falling stream of matter. This was the first observation of such a phenomenon, and provided valuable new insights into the orbital dynamics of the accreted material as it was being drawn off one star onto the companion.

Though white dwarves are highly luminous at EUV wavelengths, the brightest source in the sky is the B star of epsilon Canis Majoris, located about 600 light years away. This is an entirely different class of object though, with a photospheric temperature comparable to the cooler white dwarfs. If the ISM were really as opaque to UV radiation as was previously thought, the EUV radiation at 60 nm would have been attenuated by at least 10^{35}! Since this was patently not the case, astronomers concluded that the transparent ISM region in that direction had to extend to at least 600 light years, assuming a mean neutral hydrogen density of only 0.002 atoms per cubic cm. Astronomer Joe Cassinelli (University of Wisconsin), captured spectra of the photospheric continuum radiation at long wavelengths, which differed significantly from that predicted by theory, with the recorded emission lines being derived from hot, shocked winds blowing out from the star at shorter wavelengths.

Beyond the Milky Way, at a distance of more than a billion light years away, the BL Lacertae object PKS 2155-304 was shown to radiate profusely across the spectrum from the visible to hard X-rays. These prodigious emissions, which exceed that of the entre Milky Way Galaxy, appeared to originate inside a minute structure with dimensions no bigger than the size of our Solar System, exhibiting intensity variations within hours. As we have seen, EUVE has observed as many as 22 extra-galactic sources— including AGNS, BL Lacertae objects and Seyfert galaxies.

The ratio of AGNs to BL Lacertae objects also differed from theoretical expectations based on X-ray observations alone. This led Tim Carone (University of California, Berkeley) and his colleagues to conclude that EUV-absorbing neutral gas must be present in AGNs but not in BL Lacertae objects.

In our own backyard, EUVE has provided fascinating new insights into the Jupiter-Io system, Mars and even our own Moon. Warren Moos (John Hopkins University) and his colleagues observed the torus of plasma that surrounded Jupiter at the orbital radius of Io and even managed to resolve at least a dozen emission lines from sulfur and oxygen ions. The Russian scientist Vladimir Krasnopolsky employed EUVE spectra of the thin Martian atmosphere to measure helium abundances, allowing the radioactivity levels in the interior of the planet to be deduced. EUV observations of the lunar regolith provided excellent reflectance data for monitoring the solar EUV flux, because the EUVE was not designed to observe it directly. What's more, because this solar EUV flux directly influences the photochemistry of Earth's upper atmosphere, these measurements were a very valuable bi-product of the mission.

The EUVE mission was extended not once, but twice, but in the end, the usual constrainer—value for money—forced NASA to terminate the mission in 2000. EUVE satellite operations formally ended on January 31, 2001 when the spacecraft was placed in a safe hold orbit. EUVE transmitters were turned off in early February of the same year, after which the satellite re-entered the Earth's atmosphere over central Egypt at approximately 11:15 pm EST on January 30, 2002.

This brings us to the end of our exploration of the ultraviolet universe. In the next chapter, we will explore more of the high-energy cosmos that is beyond the ken of human eyes, the universe revealed by X-ray telescopes.

7. The X-Ray Universe

Scarcely a year after the end of WWII, the distinguished American astronomer Lyman Spitzer wrote up a classified report which recommended the design and construction of large space-based telescopes to cut through the turbulence of the Earth's atmosphere in order to provide the most detailed images of the universe ever produced. Indeed, he went on to approve the construction of giant telescopes up to 600 inches at a time when the 200-inch Hale reflector was still on the drawing board. Like so many great ideas, Spitzer's far-reaching vision went largely unnoticed until the advent of the Space Race a decade later. Lloyd Berkner, the then-chairman of the Space Science Board of the National Academy of Sciences, solicited more suggestions for space based astronomy projects. Initially, the response from the astronomical community was to design a number of small space telescopes capable of imaging objects at UV wavelengths. Called the Orbiting Astronomical Observatory (OAO) satellites, these consisted of a series of four American space observatories launched by NASA (with collaborations with the British Science and Research Council) during the years 1966 through 1972, and which provided the first high-quality observations of many objects at wavelengths shorter than the human eye could see. Although two of OAO missions ended in complete failure, the success of the other two greatly increased awareness within the astronomical community of the tremendous benefits of space-based observations. Indeed, it was these proof-of-concept spacecraft would eventually lead to the development of the Hubble Space Telescope (HST).

The first OAO entered the vacuum of space on April 8 1966. On board were a suite of instruments that could detect ultraviolet, X-rays and gamma rays. Before these instruments were turned on, however, the spacecraft developed a glitch with its power supply with the result that the mission was terminated after just three days. Indeed the malfunction sent the spacecraft spiraling out of control, because its solar panels could not be deployed to recharge

© Springer International Publishing Switzerland 2017
N. English, *Space Telescopes*, Astronomers' Universe,
DOI 10.1007/978-3-319-27814-8_7

its batteries. AO-2 was launched on 7 December 1968, carrying no less 11 ultraviolet telescopes! Remarkably, AO-2 remained fully operational until January 1973 and made many significant astronomical discoveries, including the uncovering of the enormous haloes of hydrogen, hundreds of thousands of kilometers across, around icy comets, as well as the UV variability of a number of novae during times when their optical brightness was fading.

OAO-B carried a 38-inch Cassegrain telescopes tailored to UV light, its mission goal to provide spectra of fainter objects than had previously been observable. Unfortunately, OAO-B's booster rocket failed to separate from the satellite following its launch on November 3 1970, and, as a result, it burned up as it re-entered the Earth's atmosphere.

OAO-3 was a success however. Launched on August 21 1972, the spacecraft proved to be the most successful of the OAO missions. Being a collaborative effort between NASA and the UK's Science Research Council (later known as the Science and Engineering Research Council), OAO-3 carried aloft an X-ray detector built by University College London's Mullard Space Science Laboratory in addition to a 80 cm UV telescope designed by a team of engineers from Princeton University. After its launch, it was named Copernicus to mark the 500th anniversary of the birth of the Polish astronomer, Nicolaus Copernicus. The main body of Copernicus measured 3×2 meters. Its solar panels were fixed at an angle of 34 degrees to the observing axis, and were maintained within 30 degrees of the Sun. This restriction meant that only limited patches of the sky could be monitored at certain parts of the year. The astronomical instruments, which were also co-aligned with the spacecraft, consisted of a UV telescope residing in the central cylinder of the satellite, as well as an X-ray experiment. While the UV telescope was observing, the X-ray detectors would take background measurements and every now and then the X-ray detector observed an X-ray source in the same field of view of the UV target (Fig. 7.1).

Before the Copernicus satellite was launched into space to begin its mission, cosmologists were still in the dark about one of the key predictions of the Big Bang theory. This pertained to the heavy isotope of deuterium, which was only detected on Earth but had not yet been found in interstellar space. But theory had predicted that deuterium would have been formed within the first

Fɪɢ. 7.1 A replica of the grazing incidence X-ray telescope on board Copernicus (Image credit: University College London)

three minutes after the initial expansion of the universe. What is more, cosmologists had predicted the exact amount of deuterium was strongly dependent upon the initial density of the universe. Specifically, if the deuterium abundance was less than one part per million (by mass) then the space-time geometry of the universe was closed. On the other hand, if it was about one part per million it would be flat and if more than one part per million then the cosmos had an open geometry.

Copernicus soon found what it was looking for with the detection of the spectral signature of deuterium superimposed upon the UV spectrum of the star Beta Centauri. That said, the levels it calculated—some 15 parts per billion—suggested that the geometry of space-time must be open. Some scientists were quick to point out that other processes—in particular supernovae and quasars—could have added to the primordial deuterium—making it more difficult to determine if our universe is geometrically flat, open or closed.

Copernicus also discovered extremely hot, ionized gas in the interstellar medium with a temperature upwards of 200,000 K. These included super-heated filaments of gas never before seen emanating from active star forming regions, such as the Orion Molecular Cloud Complex (Fig. 7.2).

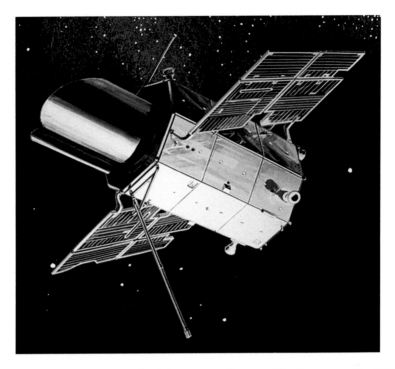

Fig. 7.2 Artist's impression of NASA's Copernicus satellite (Image credit: NASA)

The British-made X-ray instrumentation on board Copernicus also yielded valuable new data on some of the universe's most enigmatic objects. For example, it confirmed the 4.8 hour period of X-ray pulsations produced by a source known as Cygnus X-3, which was later confirmed to be a close binary system in which the matter from a dwarf star was being accreted by a neutron star. The detection of highly ionized forms of iron in its spectrum indicated that the material which had been accreted had been heated to extremely high temperatures—of the order of 10 million Kelvin!

Copernicus continued returning high resolution spectra of hundreds of stars, as well as making significant discoveries of several long-period pulsars, with rotation times of many minutes instead of the more typical millisecond variety. The spacecraft ended its life in February 1981.

Other OAOs followed fast on the heels of the highly successful Copernicus mission but were not as long-lived. O-7 was launched on September 29, 1971 and remained in operation until July 9, 1974. Like the other Orbiting Solar Observatory missions, it was primarily a solar observatory designed to point a battery of UV and X-ray telescopes at the Sun from a platform mounted on a cylindrical wheel. The detectors for observing cosmic X-ray sources were the X-ray proportional counters, built by MIT, the hard X-ray telescope by UC San Diego, and the Gamma Ray Monitor by the University of New Hampshire.

Uhuru

In 1963, Ricardo Giacconi had proposed an extensive X-ray astronomy research program to NASA that included a state-of-the-art X-ray telescope with focusable Wolter-type optics. He suggested that this should follow a small, dedicated X-ray spacecraft that had conventional, non-focusing sensors. NASA eventually agreed to the latter in 1965, and this was to become the Uhuru spacecraft launched in 1970. The 140 kilogram SAS-1 (Uhuru), was placed in a 550 kilometer circular orbit after it was launched by an American Scout rocket from the Italian San Marco platform off the coast of Kenya. As this platform was only 3 degrees south of the equator, it allowed the Scout to put the spacecraft into a near-equatorial orbit using the minimum of propellant (Fig. 7.3).

Uhuru carried a total of 65 kilos of payload, which included two X-ray telescopes, which were mounted back to back and which scanned the sky as the spacecraft spun about its axis at a rate of about 30 degrees per minute. The collimators of the X-ray telescopes yielded fields of view of 5×5 degrees and 0.5×5 degrees. Its detectors were argon-filled proportional counters sensitive to X-rays in the range from 2 to 20 keV. During the early phase of operation, the spin axis of the spacecraft was adjusted every 24 hours to scan a different 10 degree-wide band of sky. This strategy permitted the whole sky to be observed about once every two months.

All areas of the sky were not scanned with equal frequency, however, as the scan-patterns were arranged so that the Crab Nebula and the region of the center of the Milky Way near ScoX-1 could be observed most frequently. Strong X-ray sources were shown by Uhuru to be concentrated within a region of about one arc minute wide, and weaker ones about 15 arc minutes wide.

FIG. 7.3 The Uhuru X-ray satellite (Image credit: NASA)

Data was transmitted in both real time and following recording by a tape recorder, during a data dump once per orbit to the ground station at Quito, Ecuador. Unfortunately, six weeks after launch the tape recorder stopped working, and NASA was forced to bring other receiving stations on-line to pick up as much data as possible in real time, resulting in about 50 % of the data being received (Fig. 7.4).

Uhuru was able to detect weaker X-ray sources than any of the sounding rockets that came before it. In addition, Uhuru was able to measure the positions of X-ray sources with much greater accuracy. As a result, over its two-year mission lifetime Uhuru identified a total of 339 discrete X-ray sources, 100 of which had their locations pinpointed accurately enough to enable their visible and radio counterparts to be identified. Unlike stars at optical wavelengths, Uhuru found that X-ray sources tended to be highly variable in their intensity. Uhuru helped solve a number of problems. For example, Centaurus X-3 had been discovered by a team from Lawrence Livermore Laboratory in 1967, but later attempts by Ken Pounds and his team at Leicester University in the UK failed to detect any X-ray emissions from it. At first, it was doubted by many

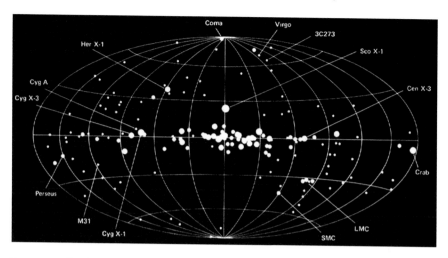

FIG. 7.4 All-sky X-ray map captured by the Uhuru X-ray satellite. Note how most sources are located near the Galactic equator (Image credit: NASA)

astronomers that Centaurus X-3 was a real object at all. But a short time after launch, Uhuru showed that Centaurus X-3 certainly did exist. It was the extreme variability of the source that explained why it was not detected in previous searches. Uhuru measured a regular periodicity of 4.84 seconds for Cygnus X-3, showing from its short period that it was most likely a rotating neutron star. Further, more detailed studies uncovered another pattern within its pulsation; it also varied slightly with a period of 2.09 days. Finally, the X-rays disappeared completely for 11 hours every 2.09 days, indicating that Centaurus X-3 was part of an eclipsing binary system. This was confirmed when a blue supergiant of 20 solar masses at Centaurus X-3's location was imaged by ground based telescopes that confirmed the 2.09 day binary variations.

It wasn't long before Uhuru was instructed to look for other eclipsing binaries, and they were rewarded in late 1971 when Giaconni's AS &E group detected periods of 1.24 seconds and 1.7 days in the X-ray emissions of Hercules X-1. Again these pulses were found to be Doppler-shifted with a period of 1.7 days. The visible counterpart was found to be a blue star, HZ Herculis, of two solar masses, which also exhibited variations at the binary period. But there was more to it than that, as when its optical emissions reached their peak, the spectrum was shown to be that of a B8 star, but at minimum it had changed to that of the cooler F0 star. This bizarre optical variability of HZ Herculis had been known for some time, but the discovery of its companion was a pulsating neutron star provided the explanation that astronomers needed.

X-rays from the neutron star Hercules X-1 were apparently heating up one side of HZ Herculis to 20,000 K. So as the system revolves every 1.7 days astronomers were observing the hot and cool sides of HZ Herculis, respectively, explaining its peculiar spectral changes. What is more, Hercules X-1's X-ray emission was only observed for about 12 days in every 35, displayed in two periods of about nine and three days in duration. During the nine day period Uhuru detected a single, strong X-ray pulse every 1.24 seconds, but during the three day period it was not detectable. As mission investigator Kenneth Brecher of MIT pointed out, the X-ray source still seemed to be heating up its optical companion, even when the X-ray source was not detectable from Earth. As time went on, researchers realized that the orbit of the binary was actually precessing with a period of 35 days, such that one of the two X-ray beams from the poles of the neutron star was intercepting the Earth over a nine day period, whilst both these X-ray jets were beamed along our line of sight over the three day period.

Uhuru also through considerable light on the nature of the variable source Cygnus X-1, which was first identified as an X-ray source by an NRL rocket experiment back in 1964. Measurements conducted shortly afterwards in 1965 showed that its intensity had decreased by 75 % from its 1964 level. Further measurements over the next five years confirmed that Cygnus X-1's X-ray intensity varied over timescales of months. Indeed, it was one of the first objects to be measured by Uhuru, being observed on 21st and 27th December 1970 and again on 4th January 1971. The results baffled astronomers though, as they seemed to show intermittent pulsations at variable frequencies, with the frequency of the pulses varying in ways that were, at best, erratic. In addition, the X-ray source was shown to flicker at the fastest rate measurable by Uhuru. Other astronomers began to look at Cygnus X-1 at optical wavelengths in order to gain further information to unravel what type of source it was, but Uhuru's position estimate was not sufficiently accurate to allow its optical counterpart to be identified unambiguously.

Shortly after the launch of Uhuru spacecraft, the MIT group observed Cygnus X-1 using a rocket experiment which had a maximum time resolution of 1 millisecond, compared with the 96 millisecond resolution of Uhuru. The flickering of Cygnus X-1 could just about be resolved by this rocket experiment, implying that the source had to be 300 km across, or one light millisecond. What is more, the same sounding rocket experiment was able to narrow down the position of Cygnus X-1 with sufficient accuracy to prompt Paul Murdin and Louis Webster of the Royal Greenwich Observatory to suggest that its optical counterpart was the blue supergiant HD 226868. This was followed up by an analysis of the spectrum of this blue supergiant, which found that it had a 5.6 day periodicity, strongly suggesting that Cygnus X-1 and HD 226868 were binary star systems. Moreover, the masses of the blue supergiant and Cygnus X-1, which weighed in at 30 and 9 solar masses, respectively, were too heavy to be legitimate neutron stars. The simplest and best explanation was that a black hole must exist here! But accepting the latter hypothesis needed more scientific evidence before it could be seriously entertained.

Whilst the optical observations of HD 226868 were ongoing, astronomers enjoyed a stroke of luck that increased their confidence that the blue supergiant was indeed the optical counterpart of Cygnus X-1. The first piece of evidence came from Braes and Miley, who found a radio source within one arc second of HD 226868 using the Westerbork radio interferometer based in Holland. Then, over a period of three months from March through May 1971, these radio emissions increased in intensity by at least a factor of four. At the same time, the X-ray intensity of Cygnus X-1 measured over the energy range between 2 and 6 keV was found by Uhuru to decrease by a factor of three. It now became clear that Cygnus X-1, HD226868 and the radio source were one and the same system. So Cygnus X-1 really did appear to be a black hole. Further evidence in support of its binary nature came in 1976, when S.S. Holt and colleagues found a 5.6 day modulation in the X-ray intensity of Cygnus X-1 using the Ariel 5 spacecraft, confirming the existence of the blue supergiant HD 226868.

Thus, within a year of its launch Uhuru had helped to show that Centaurus X-3 and Hercules X-1 were both rapidly rotating neutron stars, each in a binary system with a blue star.

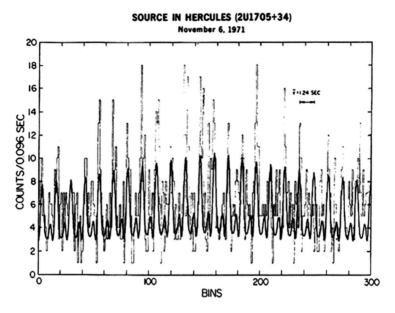

FIG. 7.5 Uhuru observations revealed the presence of X-ray pulsations in Her X-1 (1.2 s) and confirmed that it contains a rapidly rotating neutron star. Figure adapted from figures by E. Schreier, STScI, taken from Fig. 7-2a in Charles and Seward (Image credit: NASA)

Further observations revealed that the spin rate of these two neutron stars was not slowing down like the Crab pulsar, but actually speeding up! So rather than losing mass, these two neuron stars appeared to be gaining it. Applying the conservation of angular momentum to these systems, theoreticians surmised that the binary companion was losing mass to the neutron star by means of a spinning accretion disk around the neutron star (Fig. 7.5).

The hot plasma generated in the accretion disk would be directed by the neutron star's intense magnetic field onto its surface as well as its magnetic poles. This would produce a number of X-ray hot spots, fuelled by the infalling plasma. The same was true with Cygnus X-1, except for the notable difference that Cygnus X-1 is a black hole and not a neutron star (Fig. 7.6).

An accretion disk would still form around the black hole, but X-rays would be produced as the inner edge of the accretion disk loses mass to the black hole. We would not see regularly pulsed X-rays, however, because unlike the case of a neutron star, the black hole would not have a magnetic axis misaligned with the spin axis.

FIG. 7.6 An artist's rendition of a black hole and its accretion disk. Image credit: © NASA/Dana Berry, SkyWorks Digital (Image credit: NASA)

Uhuru contributed to many other areas of astronomy. It measured X-ray emission from a number of globular clusters, the Large and Small Magellanic Clouds (LMC and SMC, respectively), some spiral galaxies in the Local Group, including M31, the radio galaxy Centaurus A, some Seyfert galaxies (including NGC 4151), and some clusters of galaxies (including the Virgo cluster). Uhuru found the SMC to be dominated in X-rays by the source SMC X-1 or 3U 0115-73, which was found to be the X-ray binary companion of the blue supergiant Sanduleak 160, but the four sources in the LMC could not be clearly correlated with any objects in other wavebands. Interestingly, the X-ray emission from a number of clusters of galaxies seemed to be greater than would be expected from the sum of its individual galaxies. Unfortunately, Uhuru's angular resolution was not good enough to pinpoint the location of X-ray sources in galaxies or galactic clusters, however. To carry that out astronomers would have to build a more powerful X-ray telescope.

As the first dedicated, non-solar, X-ray observatory, the legacy of Uhuru ushered in the first era of X-ray astronomy and the start of a new one. No longer did experimenters have to rely on just a few minutes of observational data from a sounding rocket.

Now astronomers could monitor the same objects over long durations with Uhuru's superior technology, much in the same way that they had used ground-based telescopes over the centuries. This enabled them to measure the variations of X-ray intensities in timescales ranging from 0.1 seconds to many months and, in the process, begin to understand the causes of some of those variations.

As soon as the first non-solar X-ray source had been discovered in 1962, the wider universe had looked a relatively quiet and predictable place. Every now and then a bright nova or supernova would disturb this otherwise quiescent environment. But within a decade as more and more X-ray sources were identified, our view of the cosmos changed forever as a number of highly variable, highly energetic sources, such as neutron stars, black holes and pulsars. Quite literally, the universe became a much more interesting place.

The elucidation of the nature of Cygnus X-1 required the collation of data from various spacecraft, sounding rockets, and ground based optical and radio observatories. Astronomy had truly become an integrated discipline by the early 1970s, as scientists began to realize that they could use any waveband and any ground or space type of observatory to solve an outstanding problem in astrophysics. Indeed, it could be said that this integrated approach had already begun with radio astronomy in 1963, after which time radio and optical observations had both been employed to help explain the nature of quasars. This new approach was the wave of the future.

After the discovery of the first non-solar X-ray sources back in 1962, new discoveries came flooding in, leading to a large increase in the number of X-ray papers published, increasing from a single one in 1962 more than 300 scarcely a decade later. This was paralleled by a significant increase in the percentage of American scientists working in the field of X-ray astronomy, increasing from 0.8 % in 1962 to 11.2 % by 1972. Such a rapid increase in X-ray related research made it the third largest field outside optical and radio astronomy. The X-ray universe had come of age.

In the years after Uhuru, pioneers in the field of X-ray astronomy, including Riccardo Giacconi, continued to lobby NASA and other space agencies for a spacecraft equipped with a new, state-of-the-art telescope to follow it up. Unfortunately his suggestions received rather scant support from ether NASA or the wider astronomical community in general. That being said, as the case for a new X-ray observatory strengthened, NASA eventually agreed to fund his team to build a new telescope to carry on ground-breaking solar research. One such telescope was to fly on a Skylab mission scheduled for launch in 1973, so Giacconi set to work developing the design and manufacturing techniques needed for these new solar programs. He also continued to canvass support for the development of focusing X-ray optics for a separate, non-solar program.

In 1967 NASA created the Astronomy Missions Board of distinguished scientists to act as an advisory body for possible future astronomy missions. The board consulted widely and eventually recommended a greatly expanded program of high energy astronomy, with priority to be given to a number of Large High Energy Astronomy Observatories (HEAOs), one of which would be devoted to X-ray research. A feasibility study was then undertaken into the design of these HEAO missions, and this resulted in the proposal of an enormous 11-ton spacecraft. The next stage of the project involved lobbying support from NASA, the OMB and Congress, and in May 1970 competitive preliminary spacecraft design studies were left to Grumman and TRW.

While these industrial studies were ongoing, NASA issued an Announcement of Opportunity calling for experiment proposals, and after carefully assessing all the responses NASA ruled in favor of a program involving four HEAO spacecraft, two of which, were to be slowly rotating survey spacecraft, with two being fully pointable. The first two were to be designed to discover faint new X-ray sources, to measure their positions accurately, and to measure their spectral properties. These two spacecraft were also expected to carry gamma ray and cosmic ray instruments. The second two HEAOs, which were pointable, were to de designed to enable short-period variations to be measured in the X-ray intensities of known sources. These second two spacecraft would also have focusing X-ray optics to provide X-ray imaging, and to enable position measurements to be made with very high accuracy.

In 1972 a committee of eminent astronomers published a study commissioned by the Academy of Sciences, on the feasibility of both ground- and space-based facilities required for the decade ahead. Chaired by Caltech astronomer, Jesse Greenstein, the report strongly supported the four spacecraft HEAO programs, but no sooner was the ink dry on their report that NASA decided to jettison the fourth spacecraft as a cost-cutting measure. This was followed in January 1973 by another announcement by NASA that they were to cut $200 million off the budget for the HEAO program. As a result, all significant work on the HEAO program had to be suspended while these budget cuts were being executed. Morale among astronomers was understandably low, with the result that many were concerned that the program was about to be mothballed.

The Council of the American Astronomical Society were quick to condemn the NASA announcement, with an equally rapid retort coming from NASA, recommending that the mass of each spacecraft should be reduced from 11 to just 3 tons. This drastic mass reduction would allow the original Titan III launcher to be replaced by the much less expensive Atlas-Centaur, resulting in a cost reduction for the total program of 70 %. After much deliberation, this was the program that was finally approved.

In retrospect, it seems incredible that NASA had originally proposed launching four 11-ton HEAO spacecraft in as many years, when one considers that the mass of the Hubble Space Telescope is only 9 tons. But to be fair, the HEAO program had been conceptualized during the late 1960s when NASA still entertained ambitious plans, including the development of a reusable space shuttle, a base in lunar orbit in 1976, a twelve-man space station by 1975 and even a 100-man space station by 1985. The space shuttle program got the green light in 1972, but the financial crisis over the next few years served only to highlight the unfeasibility of many of these programs, at least for the immediate future. The newly approved program was to consist of three 3-ton spacecraft; HEAO-1 was to be scanning mission with four X-ray instruments, HEAO-2 was to be an X-ray imaging mission based on Giacconi's focusing X–ray telescope, and the HEAO-3 was to be a gamma-and cosmic-ray mission. These spacecraft were successfully launched in August 1977, November 1978, and September 1979, respectively.

HEAO-1

The HEAO-1 spacecraft with its non-focusing X-ray detectors was the last spacecraft of its kind. Even though it was 5.8 meters long and weighed about 2.6 tons, HEAO-1 reached the limit of what was theoretically possible with its large proportional counters, because unlike optical telescopes, their sensitivity increases only slowly with detector area. HEAO-1's carried four experiments, the largest of which was NRL's Large Area X-ray Survey Experiment, consisting of seven proportional counters, six on one face and the remaining one on the opposite side of the spacecraft. These were sensitive to X-ray energies over the range 0.25 to 20 keV and designed to detect very weak sources, as well as to monitor rapid fluctuations of likely black hole candidates such as Cygnus X-1. A second experiment, consisting of six proportional counters, covering the range 0.2 to 60 keV, was designed to measure the X-ray background radiation in an effort to elucidate its origin. The position of 1 to 15 keV sources was measured by the Scanning Modulation Collimator Experiment to an accuracy of about 5 to 30 arc seconds. This was an order of magnitude better than that measured by Uhuru. Finally, two other instruments—the Hard X-ray and Low-Energy Gamma-ray Experiment—were designed to detect and measure X-ray sources and their associated spectra in the energy range from 10 keV to 10 MeV.

HEAO-1 orbited once every 33 minutes, allowing its instruments to scan the entire sky carefully and thoroughly. Every 12 hours the spin axis was moved by 0.5 degrees to enable the spacecraft to complete an all sky survey in just six months. After the first survey was completed a second one was initiated, complemented with a number of pointed observations of particular sources of interest to astronomers. Although HEAO-1 mission was designed for a one-year lifetime, it continued to operate for another five months before its attitude control gas was finally exhausted.

Whilst HEAO-1 did not produce the spectacular images of the much superior HEAO-2, it nevertheless provided a wealth of new data, uncovering a variety of new types of X-ray sources. Not only were HEAO-1's detectors more sensitive than Uhuru's, they also reached into the very soft X-ray region below about 1 keV (or above 1 nm) where Uhuru was well-nigh blind. Within weeks of its launch HEAO-1 identified a number of intense soft X-ray sources which had previously gone undetected. Unfortunately, the position of these soft X-ray sources could only be determined to an accuracy of about 0.1 degrees, but as more and more of these sources were found, it enabled their optical counterparts to be unambiguously identified. In particular, the HEAO-1 source H0324+28 was found to be co-located with the variable star UX

Ari, a celebrated binary system of the type known as RS CVn, after the prototype in the constellation Canes Venatici.

Similar associations were then found in eight additional high-intensity, soft X-ray sources and RS CVn-type binary stars, which were only previously known to emit at radio wavelengths. Such binaries usually consist of one K- and one G-type star, with the K star being a subgiant that has already started to evolve off the main sequence of the Hertzsprung-Russell diagram. The light curves of such RS CVn systems display both the primary and secondary minima when the stars eclipse each other, as well as showing more subtle undulations as they emerge from the eclipse. These undulations and their associated radio bursts have been linked to star spots on the K subgiant, much like sunspots on our own star, the Sun, and its original coronal disturbances associated with these star spots that were thought to be the origin of the soft X-ray emissions. In 1976 the SAS-3 spacecraft had helped astronomers to understand the true character of AM Herculis, the first 'polar' to be discovered where matter is being lost from a red star and then dumped onto the magnetic poles of its white dwarf companion. What is more, the spin rate of the white dwarf was synchronized with the orbital period of the binary by the very strong magnetic field of the white dwarf.

Shortly after this discovery, the HEAO-1 satellite uncovered a binary system, prosaically named H2252-035, in which the compact object also seemed to be a white dwarf, as judged by the ratio of the X-ray to the optical brightness. In addition, optical and ultraviolet spectra lent further credence to this conclusion. Project scientists, J. Patterson and C. Price, found that the light of HD2252-035's optical counterpart, called AO Psc, pulsed every 858 seconds and also displayed a longer period modulation of 3.6 hours. These data led astronomers to believe that H2252-035 was like AM Herculis, only that the spin rate of the white dwarf is not the same as the orbital period of the binary. But this system was more peculiar than AM Herculis, however. When Patterson, N. White and F. Marshall began to measure the X-ray periodicity of H2252-035, they figured it to be 805 seconds. Clearly, the different optical and X-ray periodicities could not be attributed the spin rates of the white dwarf. This presented yet another question; what is the spin rate of the white dwarf and why did its optical and X-ray pulsation rates differ?

A decisive way forward came when astronomers realized the X-ray period is $1/16^{th}$ of the binary period, whereas the optical period is $1/15^{th}$ of the binary period. That is, every one binary period the X-ray and optical pulses are temporarily in sync with each other. The X-rays emitted by the white dwarf were evidently being absorbed and re-emitted by its companion as optical radiation. Then, as the white dwarf pops in the same sense as it orbits its companion, there are only 15 sets of optical pulses received via its companion for every 16 sets of X-ray pulses emitted by the white dwarf.

In retrospect, the model that best fits H2252-035/AO Psc is of a normal star that undergoes mass loss to the magnetic poles of its white dwarf companion via an accretion disk orbiting the white dwarf. In this particular scenario, the magnetic field of the white dwarf is not strong enough to synchronize its rate of spin to the original rate of the binary, so the system appears to be intermediate in type between the polar source AM Herculis and bona fide dwarf novae. For this reason it was called an 'intermediate polar.'

In 1979 a team led by Dr. Webster Cash, based at the University of California, began analyzing HEAO-1 sky survey data obtained from the sensors designed to detect X-ray background radiation, when they noticed what happened to be an unknown supernova remnant in the constellation of Cygnus. As they began to trace out the angular size of the source, they found that it was a remarkable 18×13 degrees in size, with an estimated temperature of about 2 million K. Curiously, the epicenter of the source was not detectable, however, although this was attributed to the X-rays being absorbed by the intervening Great Rift dust cloud, which can be observed with the naked eye as a dark band cutting across the Milky Way. Further analysis showed that thin filaments of hydrogen detected in the light of Hydrogen alpha coincided with the newly discovered ring. There are a number of massive o-type stars still in this region, so it is thought that this ring, now called the Cygnus Superbubble, was probably produced by a number of supernova explosions that occurred in a group of similar O-type stars about 3 million or so years ago (Fig. 7.7).

HEAO-1 also discovered a new candidate black hole after it discovered rapidly fluctuating X-rays, reminiscent of Cygnus X-1, from the source GX 339-4. Yet it was the satellite's discovery of the X-ray background radiation that was arguably its most lasting

Fig. 7.7 Artist's impression of the HEAO-1 satellite (Image credit: NASA)

achievement. In the first instance, the known X-ray sources as well as the galactic X-ray component had to be eliminated from the data before this background radiation could be studied. At first large swathes of the sky where the X-ray background should have been detected could not be observed owing to these masking sources, but there was still enough data remaining to enable an analysis to be conducted. The results showed that the intensity variations from the X-ray background from point to point in the sky were significantly larger than expected from normal statistical fluctuations alone, implying the real background was not truly isotropic, but was made up of many, faint unresolved sources that produced the lion's share of the background radiation.

HEAO-2: The Einstein Observatory

From the outset, HEAO-2 was revolutionary, for it the first X-ray astronomical spacecraft capable of taking images by focusing them tightly. Launched from Cape Canaveral by an Atlas-Centaur on November 13 1978, its principal instrument consisted of a

Wolter-type X-ray telescope which consisted of nested cylinders which were able to bring X-rays (in the energy range between 0.25 and 4 keV) to a sharp focus. This state-of-the-art instrument had four detectors arranged on a rotating platform. The first of these instruments; the High Resolution Imager (HRI) had a relatively wide field of view of about 0.5 degrees (one Moon diameter) and was able to produce X-ray images with a resolution of about 2 to 3 arc seconds. HRI did not have any spectral capability, however. In contrast, HEAO-2's second instrument, the Imaging Proportional Counter (IPC), was able to produce both spectral and positional information. That said, although it was considerably more sensitive to X-rays than the HRI, it exhibited poorer spatial resolution—just 1 arcminute. Spectral data could be produced on any object imaged in the telescopic field by using the Objective Grating Spectrometer (OGS) coupled to either the IPC or HRI. The best spectral data were produced by another instrument though— the so-called Solid State Spectrometer (SSS)—which had a resolution of 0.15 keV, or the Focal Plane Crystal Spectrometer (FPCS), which, although less sensitive than the SSS, had an energy resolution of about 0.001 keV. The last instrument used by HEAO-2 was the Monitor Proportional Counter (MPC), which was mounted separately from the main X-ray telescope and which was used to analyze the brightness variability and spectral characteristics of the most interesting X-ray sources.

Immediately after its launch, ground controllers experienced a nerve wracking period during which time the star sensors that were to be used to control the spacecraft, failed temporarily. Bizarrely enough, the problem was attributed to a reflection of the Moon off the Pacific Ocean! Once this had been realized, the in-orbit systems tests were completed and HEAO-2, or the Einstein Observatory, as it was came to be known, became fully operational and continued to work more or less flawlessly for two and half years until April 1981, when its attitude control system ran out of gas.

Before the advent of the Einstein Observatory, it had been widely believed that red dwarf stars were much too cool to harbor an extensive corona and so ought not to emit X-rays, except perhaps if they happened to be flare stars during outbursts. But here's where Einstein produced one of its earliest surprises, when it showed that normal red dwarf stars, together with flare stars in their quiescent phase, were very strong X-ray emitters. Indeed, they were found to emit an astonishing 10 % of their total energy at X-ray wavelengths. This had to mean that they had hot coronas like the Sun. Another surprise came when the Einstein observatory observed young, hot stars. It had been thought that these stars—mostly O, B and A stars—would not have hot coronas. That was because these stars do not apparently have a turbulent convective zone just below their surfaces, which was traditionally believed to be the prime source of coronal heating. Here also, HEAO-2 proved the theorists to be mistaken, as some of these stars were also found to be significant X-ray emitters, with the only exceptions being the very cool giants and super-giant stars. Moreover, within any one stellar population, the X-ray intensities covered an enormous range- up to three orders of magnitude— which caused another surprise as stars of this kind should have similar coronal properties. Clearly the theories of stellar structure needed a radical overhaul.

These observations prompted researchers to investigate the causes of X-ray emission from red dwarfs and young, hot O-, B- and A-type stars. On first assessment, it was believed that the X-ray flux from red dwarfs might have been produced by re-connecting magnetic field loops to their highly convective atmospheres. In the case of the young, hot stars, however, it was thought that the X-rays were probably generated from shock waves produced by strong stellar winds buffeting the interstellar medium. But this also raised the question as to why ordinary stars in the mid-section of the main sequence did not exhibit such X-ray emissions?

One suggestion forwarded by some astronomers was that this could reflect differences in stellar rotation rates as a function of age. From a purely theoretical standpoint, it would be expected that fast rotating stars would, via a dynamo effect, produce stronger magnetic fields than more slowly rotating stars, and that these stronger magnetic fields would in turn retain more plasma and thus generate more X-rays. It was known that older stars popped faster than older stars and so Zoleinski, Stren, Antiochus and Underwood used Einstein observations of stars in the relatively young Hyades cluster to see how their X-ray emission compared with that of other main sequence stars. The Hyades cluster was known to be about 600 million years old, compared with the Sun's

4.5 billion year-old age, and Zoleinski and colleagues found that on average, solar-type stars in the cluster were about 50 times more luminous in X-rays than the Sun. Astronomer, Jean Pierre Cauillat also found that the average X-ray luminosity of stars in the even younger Pleiades cluster was about ten times greater than those in the Hyades, thus adding to credibility to the theory linking X-ray emission to age-related stellar rotation rates.

In addition to this work, the Einstein Observatory provided much new data on sources called X-ray transients. In particular, it helped scientists to understand the structure of the transient AO538-66, which was first uncovered back in 1977 by Nick White of MSSL and Geoff Carpenter of Birmingham University, using the Ariel 5 satellite. Unfortunately, Ariel 5's poor spatial resolution meant that the X-ray source could not be located accurately enough for an optical identification, but it was fairly clear that it was in the direction of the large Magellanic cloud. The Ariel 5 data also showed that not only was A0583-66 highly variable, but it had a regular 16 day period between peaks, indicating that it was a binary. Mark Johnson (MIT), Richard Griffiths (Harvard) and Marti Ward (Cambridge) then used HEAO-1 to get a more accurate location of the X-ray source, which they found coincided with a variable star. Its optical period of 16.6 days was deduced by Gerry Skinner of Birmingham University using old Harvard Observatory plates, but he also found that there had to be periods in the preceding 50 years when the variability had disappeared. This strangely variable optical source was clearly the X-ray source's binary companion.

In 1980, astronomers Philip Charles of Oxford University and John Thorstensen of Dartmouth College obtained the optical spectrum of A0538-66 revealing that of a typical B-type star, with hydrogen and helium absorption lines. These red shifted emission lines and blue shifted absorption lines suggested that the star was surrounded by a shell of gas hurtling away from AO538-66 at velocities in excess of 3000 km/s. Furthermore, during its quiescent phase its spectrum allowed it to be definitively co-located with the Large Magellanic Cloud (LMC), as it was receding from the Solar System at the same velocity as the LMC itself (Fig. 7.8).

The new data garnered by Einstein allowed scientists to estimate A0538-66's absolute X-ray luminosity to be about 10^{32} Watts at its peak, making it the most luminous X-ray stellar source

Fig. 7.8 Artist's impression of the Einstein Observatory in Earth orbit (Image credit: www.Harvard.edu)

known at the time. Einstein project scientists, Gerry Skinner and Martin Weisskopf, of the Marshall Space Center, and their collaborators then detected X-ray pulsations with a period of 69 milliseconds, indicating that the X-ray source was a rapidly rotating neutron star. What is more, observations carried out by the IUE spacecraft showed that the optical source is redder when it is brighter. Under normal circumstances, the binary companion of an X-ray star becomes hotter and bluer when it is exposed to X-rays, but this was not the case with this system. The reason appeared to be that, at the beginning of the outburst, the optical source covered a much larger area, but as a consequence appeared cooler and thus redder than normal.

Spectral lines undergo broadening if their source is rotating rapidly, probably close to the fragmentation speed of the star. So the hypothetical model of the system has an optical source, with a large equatorial disk of material, in a binary system with the X-ray source A0538-66. The orbit of the latter is highly eccentric and once every orbit it passes though the disk of the optical source. This creates a burst of X-ray energy that heats and partially depletes the disk, as its material is accreted onto the surface of the neutron star at its poles. After a number of orbits the material in this disk is not dense enough to overcome the X-ray star's magnetosphere, which expands as the disc pressure reduces, and the active period of the binary ends. The optical star continues to emit material, however, and eventually the disk accumulates enough for the process to start again.

For the first time, in addition to observing transients, the Einstein Observatory was able to image supernova remnants (SNRs). This showed, for example, that the X-ray images of the Crab Nebula are much smaller than the Nebula in visible light. The pulsar famously associated with the Crab, which the spacecraft also imagined, was plainly shown to be off center compared with the nebula imaged in X-rays. Einstein also imaged the Tycho and Cas A supernova remnants, and confirmed that their shell structure in X-rays was due to their exceptionally high temperature. Both Tycho and Cas A exhibited outer shells with temperatures of the order of 40 to 50 million K, caused by the expanding shells' shock-induced heating of the interstellar gas, but, intriguingly, the Cas A remnant also showed an inner shell of slightly cooler, denser gas (Fig. 7.9).

The outer shell was shown to be decelerated by pockets of interstellar gas, while the inner shell was shown to be catching up and beginning to interact with it, setting the inner shell aglow in X-rays also.

The Puppis SNR was analyzed by Frank Winkler, Claude Canizares and colleagues using the FPFS on Einstein with its hitherto unprecedented resolution in X-rays (1 eV). They identified highly ionized emission lines of neon, iron and oxygen which enabled the astronomers to quantify the amounts of these elements present. The data revealed that Puppis A had a much larger amount of oxygen and neon than one would expect under normal circumstances, indicating that the progenitor star must have been at least 25 solar masses and which ended its short life in a cataclysmic type II supernova explosion. The current temperature of the gas was estimated to be somewhere between 2 and 5 million Kelvin.

FIG. 7.9 The false color X-ray image of the Tycho supernova remnant imaged by the Einstein Observatory (Image credit: NASA)

Prior to the successful launch of the Einstein Observatory, the Vela and Crab remnants were known to contain pulsars that also emit X-rays. What is more, only the Crab source was known to actually pulse in X-rays. Einstein quickly unveiled three more X-ray pulsars in SNRs, strengthening the link between SNRs and X-ray pulsars. One of these SNRs, prosaically named 0540-693, which was discovered by Rick Hardnen, David Helfand and Frederick Seward in the LMC, was shown to be remarkably similar to the Crab Nebula, exhibiting blue nebula emission due to synchrotron radiation. Its associated pulsar, which was exceedingly bright, had a measured period of 50 milliseconds, a little larger than the 33 milliseconds for the Crab, and it was shown to pulsate at both visible and X-rays wavelengths. Thus, the Crab Nebula is not unique.

When astronomers trained their large telescopes on the radio galaxy Centaurus A it was seen as a vast storm of stars some 6 arc minutes in angular diameter crossed by a prominent dust lane. At Centaurus A's distance, 6 arc minutes corresponds to about an actual diameter of 15,600 light years. Sophisticated image processing techniques have shown that the galaxy extends to about 30 arcminutes in visible light, but in radio waves it was found to stretch an astonishing 9 degrees across the sky in an orientation perpendicular to the dust lane.

In 1973 the OSO-7 satellite uncovered solid evidence that Centaurus A is an X-ray source, varying in intensity over a number of days. Subsequent observations made by both the Ariel 5 and OSO-8 spacecraft detected X-ray variations over less than one day. This suggested to astronomers that the X-ray source within Centaurus A must be less than one light day in diameter, compared with the 75,000 light year diameter optical source, and the even larger (1.25 million light year) radio source. Sometime later, the SAS-3 spacecraft confirmed that the X-ray source was indeed relatively tiny, being less than one arc second in diameter, located within a few arc seconds of the center of Centaurus A.

The Einstein Observatory was also the first to image the X-ray source of Centaurus A, showing that most of the energy emanated from a very small region at the Galactic center. Furthermore, it imaged an enormous jet emitting X-rays offset slightly from the nucleus, with an orientation similar to that of the radio image of Centaurus A, except that it was very much smaller. Although the X-ray source was very small, its luminosity was found to be about 10^{35} Watts, which is about five orders of magnitude greater than that of the center of the great spiral galaxy in Andromeda(M31). Getting an intense X-ray signature from such a small source provided good indirect evidence that that there must be a supermassive black hole at the center of Centaurus A.

As far back as the 1960s, another radio galaxy, M87, also known as Virgo A or 3C 274, had also been found to be an X-ray source. Indeed, a small optical jet had been observed emerging from the nucleus of M87 as long ago as 1917, but it was all but forgotten until it was rediscovered by Walter Baade and Rudolph Minkowski in 1954. What is more, the jet was highly polarized suggesting that it was being generated by synchrotron radiation from high energy electrons spiraling along the galaxy's strong magnetic field. In addition, in 1966 a small optical counter-jet was found by the distinguished American astronomer Halton Arp. Subsequent image processing of these jets showed that M87 was considerably larger than originally thought. It had a diameter of about 1 million light years, some ten times greater than that of the Milky Way (Fig. 7.10).

Fɪɢ. **7.10** Hubble optical light image of the jet streaming the center of M87 (Image credit: NASA)

When the Einstein Observatory was first turned on M87, it clearly showed that both the nucleus and its associated jet were strong emitters of X-rays. Meanwhile images captured by large, ground-based optical telescopes forced astronomers to conclude, from the density of M87's core and the velocity distributions of its constituent stars, that there must be about 5 billion solar masses of material packed into a volume only ten light years in diameter. This indicated the presence of a supermassive black hole there. If that were the case however, the X-rays emissions from the nucleus of this galaxy would be higher than that observed by Einstein. On the other hand, the existence of the jet, which was imaged at radio, optical and X-ray wavelengths, strongly indicated the presence of a large accretion disk immediately around the galactic nucleus. Furthermore, this accretion disk would serve to focus the jet along its axis. That said, in these early years, there were still many in the astronomical community who remained skeptical that there really existed a supermassive black hole at the center of this galaxy.

Astronomers D. Fabricant and P. Gorenstein analyzed the X-ray emissions from M87, as measured by the Einstein Observatory in 1983. These observations extended beyond the edge of the visible light image of the galaxy. Astonishingly, the data suggested that there was ten times more mass in the central region of M87 that was completely invisible to even our largest optical telescopes. What's more, when extrapolated to the entire galaxy, the astronomers suggested that the visible stuff amounted to just 0.5 % of its total mass! What is more, this 'dark matter' density increased with distance from the center. Its enormous mass also suggested why M87 was seen to be stationary at the hub of the Virgo cluster, with its other constituent galaxies orbiting it at velocities up to 1500 km/s.

The quasar 3C 273 had been found to be an intense X-ray source back in 1967 and ten years later the Cos-B spacecraft contributed a discovery of its own, showing that it was also an intense source of gamma rays. Then in 1979, using the superior resolution of the Einstein Observatory astronomers found that the X-ray emissions from 3C 273 were about a million times that of the Milky Way, and varying over periods as short as half a day. Thus, astronomers reasoned that the underlying source had to be emitted from a region no larger than the diameter of our own *Solar System*. This provided further observational evidence for the reality of a supermassive black hole. Furthermore, this idea was substantiated by further observations at visible wavelengths which showed a jet apparently coming from the center of 3C 273. Further observations at radio and X-ray wavebands captured the same jet. So the galaxies Centaurus A, Virgo A (M87 and 3C 274) and 3C 273 all seemed to contain a supermassive black hole at their center and jets extending deep into intergalactic space seem to be one of their characteristic features.

The Einstein Observatory undertook extensive deep sky surveys over the 0.3 to 3.5 keV energy range in order to ascertain whether the X-ray background radiation could be co-located with discrete sources. In the first of these deep sky surveys, Einstein employed the IPC and HRI instruments to image eight fields at high galactic latitudes, well away from any of the known X-ray sources.

Using exposure times lasting up to 24 hours, Einstein discovered over 100 new sources. Of these just 25 were selected for more detailed analysis. Intriguingly, none of these were found to be quasars, many came from individual stars, one was a cluster of galaxies and another source was identified in an otherwise normal galaxy. Seven sources could not be clearly identified and very likely extragalactic.

In the second survey conducted by Einstein, which became known as the medium sensitivity survey (MSS), a variety of unknown objects that were first detected in normal IPC frames were re-analyzed by the astronomers Gioai, Maccacaro, Stocke and colleagues. The MSS yielded 800 or so new objects, 53 % of which were found to be quasars and active galactic nuclei (AGN), 27 % were stellar in origins, 13 % were clusters of galaxies, 5 % were BL Lac objects, and 2 % were normal galaxies. Comparing these findings to the less deep survey conducted by HEAO-1, 51 % of which were clusters of galaxies, so with successively deeper surveys the cluster figures reduced from 51 % (HEAO-1) to 13 % (MSS) to 6 % (DS). Collectively, these surveys showed that galactic clusters didn't seem to make a major contribution to the X-ray background radiation. But the opposite appeared to be true of quasars, however, where the percentage increased substantially between the HEAO-1 survey and the deeper Einstein surveys. Further detailed analysis of some hundreds of quasars measured by the Einstein Observatory led Tananbaum and colleagues to conclude that almost all of the diffuse X-ray background radiation in the range 1 to 3 keV most likely due to quasars which were simply unresolved at the time.

Arguably one of the most curious supernova remnants investigated by the Einstein Observatory was W50, first identified as a large radio source in the 1950s. In 1972 the English astronomers David Clark and James Caswell initiated a radio survey of all radio-emitting supernova remnants visible from Australia. Their studies showed that W50 had a somewhat oval shape with a point radio source nearby. Not long after the British team published their results, the Indian radio astronomer Thangasamy Velusamy discovered that they had only imaged part of the remnant. It was indeed oval in shape with the point radio source at its center, but it became increasingly clear that this point radio source was all that was left of a giant star that had exploded to produce the remnant W50.

Unbeknownst to these radio astronomers, their colleagues working at optical and X-ray wavebands were also showing interest in a point source in the area of W50. This began during the 1960s, when the Americans Bruce Stephenson and Nicholas Sanduleak had initiated a search for stars with emission line spectra, and in 1977 they published their Stephenson-Sanduleak (SS) catalog of these sources. But it was unknown to Clark, Caswell or Velusamy, with the result that the link between SS 433 in the catalog and W50's point radio source was missed. In addition, Uhuru had also detected a weak source of X-rays in the region of this point radio source, but was not picked up as anything unusual though until 1976, when the American astronomer Frederick Seward noticed that the weak Ariel 5 source nominally called A1909+04, was actually the same as the Uhuru source, and exhibited X-ray variability. Unfortunately, the location of A1909+04 was not known with sufficient accuracy for it to be correlated with any source in other wavebands, but there was good evidence to believe that it may the same as the radio source at the center of W50. The clincher was when John Shakeshaft of Cambridge University found that the point radio source flared erratically at radio wavelengths, giving more circumstantial evidence that the A1909+04 X-ray source and this point radio source were very likely one and the same object.

In subsequent attempts to find the optical counterpart of this point radio source, David Clark and David Crawford set about pinpointing its location with much greater accuracy, and this enabled Clark and Paul Murdin to locate a likely optical candidate using the 1.2 meter UK Schmidt telescope at Siding Spring Observatory, Australia. In addition, they employed the larger (3.9 m) Anglo Australian Telescope to obtain a spectrum of this optical candidate in June 1978. Their work uncovered an emission spectrum uncannily similar to that of another ex-supernova source called Cir X-1. Shakeshaft then produced an even more comprehensive position fix on the point radio source in W50, but alas it appeared to be 1 arc second from that of the original source, leaving some doubt as to whether they were really one and the same. Yet another optical spectrum was produced in July 1978. Intriguingly, some of the emission lines of the June spectrum were still recorded but, significantly, others had disappeared whilst completely new ones had been detected.

In the meantime though, Gregory and Crane had been studying radio emissions from SS stars, and by June 1978 found a highly variable radio emission from S 433. Clark instantly recognised the coordinates given in the IAU circular by Seaquist et al. as those of the point radio source in W50. The data was now clear: the optical source SS 433, the X-ray source A1909+04 and this point radio source were indeed one and the same object. But that still didn't answer the question as to its precise nature. After all, it was not that unusual to find both an X-ray and optical source that exhibited variability like this. That said, there was one curious feature of the SS 433 data—its variable spectral features—that revealed its unique (at least for the time) nature.

As we have seen previously, the first spectrum of SS 433 had been recorded on June 29 1978 by Clark and Murdin, but when the next spectrum was made taken just two weeks later, many of the features had completely changed. The mysterious nature of SS 433 led Bruce Margon of the University of California to investigate the source in September 1978. Margon found unusual emissions. What is more, during four nights in late October of the same year, Remington Stone, also based at the University of California, found that the most prominent lines on either side of the Hydrogen alpha line were drifting in frequency. This was confirmed by Augusto Mammamo and colleagues in Italy, who recorded the same effect. Normally, such frequency changes were attributed to Doppler effect, but if that were really the case, it indicated that the hydrogen gas was moving at velocities of up to 40,000 km/s, which seemed unreasonably high. Furthermore, it would also imply that the gas velocity changed by up to 1000 km/s per day, which seemed equally unlikely. In addition the gas appeared to be moving both towards and away from at the same time, since the spectral lines appeared to be simultaneously blue and red shifted.

A breakthrough came in January 1979, when Andy Fabian and Sir Martin Rees in the UK put forward the suggestion that SS 433 could be emitting not one, but two beams of gas, one directed towards and one away from us, at a varying speed. Meanwhile, independent work carried out by the Israeli astronomer Mordehai Milgrom, proposed that SS 433 possessed either a rotating pair of jets, or a ring of gas orbiting a central object, which also explained its strange spectral behavior. Significantly, Milgrom also observed both the blue- and red-shifting of the lines with a fixed period. This he attributed to the relativistic, transverse Doppler Effect. This is the red- or blue-shift predicted by special relativity that occurs when both the emitter and receiver are at the point of closest approach. At this moment in time, light emitted will be red shifted, while the light received will be blue shifted.

The next piece in the jigsaw puzzle came in early 1979, when astronomers, Bruce Margon and James Liebert, independently discovered that all the emission lines had identical, red-shifted, blue-shifted and stationary elements with a periodicity of 160 days (later revised to 164 days). The deviation of the median position of the blue- and red-shifted lines from their rest position enabled the transverse velocity of the emitting gas to be determined by George Abell and Bruce Margon to be about 80,000 km/s. This also meant that the jets of SS 433 could not have been directly pointed at or away from the Earth at any stage during its period, as the maximum velocity of approach or recession would only be 40,000 km/s. A more detailed analysis of the geometry of the SS433 pulsations indicated that the lines were tracing out a cone shaped trajectory with a half-angle of 17 degrees, and inclined an angle of 78 degrees between the axis of the cone and our line of sight. To all intents and purposes, SS433 presented as a monster spinning top with a period of 164 days! (Fig. 7.11).

Up until the late 1970s observations of SS 433 and its radio and X-ray counterparts had been carried out using data from numerous optical and radio telescopes and just one X-ray spacecraft, Ariel 5. In general, these data seemed to support the precessing jet hypothesis, but what astronomers really wanted was a good image of that jet. What's more, some astronomers still thought that the X-ray source A1909+04 was not the same as SS 433. The Einstein Observatory spacecraft resolved this uncertainty in 1979. In April of that year the HRI instrument on board Einstein measured the position of A1909+04 much more accurately than Ariel 5, confirming that it was indeed the X-ray counterpart of SS 433.

Then in October of the same year, the astronomers Seaquist, Seward and colleagues used the IPC instrument on Einstein to create a spectacular image of the jets extending some 35 arc minutes (or about 100 light years) in both directions. Curiously, the X-ray emitting jets appeared to be oriented in such a way that they

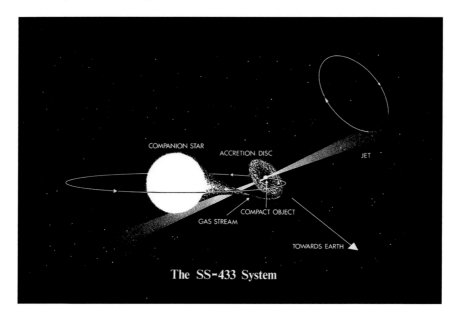

The SS-433 System

FIG. 7.11 Artist's conception of the SS 433 X-ray source (Image credit: www.youngastros.org)

were pointing straight at the lobes of the W50 shell. This suggested to researchers that W50 may not be a supernova remnant after all, but instead is a cavity blown open in the interstellar medium caused by the ejected jet material. While these new observations were being conducted, radio astronomers had been attempting to resolve SS 433 using a sophisticated technique known as interferometry. Using the partially assembled Very Large Array (VLA) Ernst Seaquist and William Gilmore carried out observations of the source at a wavelength of 6 cm and showed that SS 433 was elliptical in shape, with a semi-major axis of about 3 arc seconds. However, when they employed a longer wavelength of 20 cm, they picked up a narrow filament, aligned with the major axis at 6 cm, but extending out to 20 seconds of arc on either side of the point source. What's more, these filaments appeared to be aligned with the considerably larger X-ray jets.

As more dishes were added to the VLA during 1979 and 1980 Hjellming and Johnson began to monitor SS433 at a number of frequencies, finding that the structure of SS 433's lobe actually varied over periods of only a few weeks. Furthermore, the intensity of these radio waves was observed to increase dramatically on December 7 1979 and again on June 20 1980, generating knots of radio-emitting material on both sides of the nucleus. Further observations showed that these knots were moving, inching their way from the central source at a rate of about 9 milli-arcseconds per day. This allowed astronomers to pinpoint the distance of SS 433 to approximately 16,000 light years from Earth. Radio observations taken over the next year using the VLA, MERLIN and other radio interferometer s provided evidence that that knots were being emitted at various angles from the central source owing to its precession. These knots continued to move from the central source in a straight line, so the radio jets appeared to oscillate from side to side, rather like water coming from an oscillating water sprinkler. But what could underlie this bizarre behavior?

A team of astronomers led by David Crampton based at the Dominion Astrophysical Observatory in Canada collated several optical spectra of SS 433 over a three-month period during the summer of 1979, finding that the so-called 'stationary' lines move slightly backwards and forwards over a period of 13.1 days. The velocity curves derived from these moving spectral lines were similar to those observed for a binary containing a B-type star and a compact companion. The binary nature of the system was soon confirmed. In 1980 the Russian Anatol Cherepashchuk, and Bruce Margon working in Australia, detected both the primary and secondary minima in the light curve with a 13.1 day period. They were able to confirm that the intensities of these stationary emission lines varied over the same period.

So the emerging picture was that SS 433 was a binary system in which a large, B-type star is losing mass to a compact source via an accretion disk around the latter and having an orbital period of 13.1 days. The orbital plane of the system, as seen from Earth, was almost edge-on, so that it was possible to detect both the eclipse and the accretion disk by the B-type star and vice versa. The X-rays are thought to originate from in-falling material that is super-heated as it crashes onto the accretion disk of the compact source. Some of this material then gets ejected from the system proper, in two oppositely directed jets at right angles to the accretion disk. The plane of the accretion disk and the binary system are not the same, however, with the result that the gravitational pull from the

B-type star on the accretion disk causes both the accretion disk and the jets to precess with a period of 164 days. That said, some outstanding questions still remained unresolved. For instance, was the compact source a neutron star or a black hole? Up until this time, SS 433 was the only stellar source confirmed to have relativistic jets. Still, its X-ray intensity was unusually low for a binary system containing a compact source. What was the source of this energy and how could it cause the acceleration of the jets to the high velocities observed?

These questions were addressed by Mike Watson and his colleagues at Leicester University. They used the GSPC instrument on Exosat in the mid-1980s to measure the X-ray spectrum of SS 433's central region over the energy range 4.5 to 11 keV, which also covered the 6.7 keV spectral line of highly ionized iron. Interestingly, although they detected the change of energy of this line over the 164 day precession period, they only managed to record the blue-shifted beam, that is, the one that was directed towards us, but not its red shifted counterpart. The explanation lay with the accretion disk itself, which we observe edge-on to our line of sight. The disk must have been dense enough to block the red-shifted X-ray beam emanating from the core of SS 433.

The mass loss rate of SS 433 along with the observable beam could be calculated, assuming that the X-rays are being produced by hot gas. Then, knowing the beam's velocity, the rate of energy loss of the source along the beam could be calculated. This turned out to be about 10,000 times higher than the energy being radiated in X-rays by SS 433, lending greater credence to the idea that the true X-ray intensity of the central region of SS 433 is much higher than what had been measured. This also explained the very high speed of the jets and their unusually tight collimation. Finally, in 1991 a team led by Sandro D'Odorico employed the New Technology Telescope of the European Southern Observatory, to observe SS 433 and measure the Doppler shifts of ionized helium within the rotating accretion disk. As a consequence, his team was able to estimate the mass of the compact source to be about 0.8 solar masses and, while its companion weighed in at about 3.2 solar masses. One thing was clear though; there was far too much energy in the system for the compact source to be a white dwarf, and yet not so massive as to be a bona fide black hole. D'Odorico and colleagues were forced to conclude that the nature of the compact source was that of a neutron star.

After four successful years of active service, the Einstein Observatory satellite re-entered the Earth's atmosphere and burned up on March 25, 1982.

HEAO-3

The final High Energy Astronomy Observatory in the program— HEAO-3—was launched on September 20 1979. Incorporating a similar spacecraft 'bus' to HEAO-1 and -2, the mission of this spacecraft was dedicated to X-ray, gamma ray and cosmic ray research. It operated until May 30 1981 when its attitude control gas was depleted. Unfortunately, HEAO-3's results were far less productive than either of its predecessors, owing to the much greater difficulty of detecting and pinpointing gamma ray sources rather than X-rays.

We have already seen how the intense radio source called Sagittarius A* had been discovered at the dynamical center of the Milky Way in 1974 by the American astronomers Bruce Balik and Robert L. Brown. Sagittarius A* was estimated to have a diameter of only about 3×10^9 kilometer s (20 astronomical units), and yet within this tiny (on cosmic scales) volume it was nonetheless emitting about 10,000 times the energy of the strongest known radio pulsar. This provided good indirect evidence that there must be a supermassive black hole with a mass about 1 million times greater than the Sun at the center of our galaxy. Further evidence of its energetic nature was provided just a few years earlier, when in 1970 a balloon experiment conducted by Robert Haymes and his colleagues of Rice University captured 511 keV gamma rays emanating from the Milky Way. Seven years later a group led by Martin Leventhal of Bell Laboratories confirmed these emissions, indicating that these gamma rays were coming from the central region of the Milky Way. Theoretical studies suggested that the origin of these gamma rays were derived from photons that smashed into other particles to produce electron-positron (anti-electron) pairs, annihilating each other to generate the observed 511 keV radiation.

The gamma ray spectrometer onboard the HEAO-3 satellite was widely expected to observe these 511 keV photons too, but once activated, the instrument produced too many 'false positives' requiring two years of patient analysis by a team led by Allan Jacobson of JPL before the 511 keV line could be unambiguously detected from the central region of the Milky Way. Interestingly, the intensity seemed to vary over a period of about six months, limiting the size of the gamma ray source to about half a light year. At about the same time, the Einstein Observatory also detected X-rays from a source within about 1 arcminute of the center of the Milky Way. This varied over a period of about three years, giving an upper bound to the size of the X-ray source.

Jacobson and his team also detected 1.80 MeV gamma rays from the central region of the Milky Way using the HEAO-3 spectrometer. These gamma rays were most likely produced by the radioactive decay of Aluminium-26 to magnesium 26 with a half-life of 740,000 years. This observation was noteworthy as it implied that, as the half-life of Aluminium-26 is relatively short, it was probably still being generated in the central region of our galaxy, probably in supernovae and novae explosions.

One of the most interesting results was the detection of low energy gamma rays from the black hole candidate Cygnus X-1. HEAO-3 observed Cygnus X-1 for 170 days starting on September 27 1979 and found that it was emitting flickering X-ray emissions at about 100 keV. But it was not until the HEAO-3 data was re-analyzed in the mid-1980s by James Ling and colleagues at JPL, that low energy gamma rays were discovered. This strong gamma ray emission in the band from about 400 keV to 1.5 MeV was only present in the first two weeks of observations, however, as it then disappeared as the hard X-ray emissions increased.

Collectively the three HEAO spacecraft contributed significantly to our knowledge of the X-ray universe, finding novel objects and helping us to elucidate the high-energy processes underlying previously characterized objects. But as technology improved, astronomers began to consider even more powerful X-ray telescopes to probe the mysteries of the high energy universe. And that's where our story continues, with the legacy of the Chandra Observatory.

The Chandra X-Ray Observatory

Like all other great telescopes, the Chandra X-Ray telescope originated with a good idea. In 1976 the Chandra X-ray Observatory (called AXAF at the time) was proposed to NASA by the Italian high energy astrophysicist, Riccardo Giacconi and the American astronomer, Harvey Tananbaum. Preliminary work began the following year at Marshall Space Flight Center (MSFC) and the Smithsonian Astrophysical Observatory (SAO). In the meantime, in 1978, NASA launched the first imaging X-ray telescope, Einstein (HEAO-2), into orbit. Work continued on the AXAF project throughout the 1980s and 1990s. As a cost cutting measure, the spacecraft was redesigned in 1992. Four of the twelve planned mirrors were eliminated, as well as two of its six scientific instruments. AXAF's planned orbit was also altered to an elliptical one, reaching one third of the way to the Moon's at its farthest point. Unfortunately, this precluded the possibility of improvement or repair by the space shuttle but had the advantage of placing the spacecraft above the Earth's radiation belts for most of its orbit. AXAF was assembled and tested by TRW (now Northrop Grumman Aerospace Systems) in Redondo Beach, California.

The telescope was renamed Chandra as the result of a contest held by NASA in 1998, which drew more than 6000 submissions from all around the world. The contest winners—Jatila van der Veen and Tyrel Johnson (then a high school teacher and high school student, respectively), suggested the name in honor of the Nobel Prize–winning Indian astrophysicist Subrahmanyan Chandrasekhar (1910–1995). He is known for his work in determining the maximum mass of white dwarf stars (the so-called Chandrasekhar Limit), leading to greater understanding of high energy astronomical phenomena such as neutron stars and black holes.

Originally scheduled to be launched in December 1998, the spacecraft was delayed several months, eventually being launched in July 1999 by Space Shuttle Columbia during STS-93 under the command of a female astronaut, Eileen Collins. Weighing in at a whopping 22,753 kilograms (50,162 lb.), it was the heaviest payload ever launched by the space shuttle, a consequence of the two-stage Inertial Upper Stage booster rocket system needed to transport the spacecraft to its high, elliptical orbit extending anywhere from about 9940 miles (16,000 kilometers) to 82,650 miles (133,000 kilometers) from Earth. During its maneuvers from one target to the next, Chandra slewed more slowly than the minute hand on a clock.

Within just weeks after its successful launch, Chandra began returning data. It was operated by the SAO at the Chandra X-ray Center in Cambridge, Massachusetts, with assistance from MIT and Northrop Grumman Space Technology. The ACIS CCDs suffered particle damage during early radiation belt passages. To prevent further damage, the instrument was removed from the telescope's focal plane during these passages. The resolution of the telescope is about 0.5 arc seconds and is remarkably efficient in terms of power consumption; just 2 kilowatts or about the same as a domestic hairdryer! Data from Chandra are transmitted via the Deep Space Network stations to the Jet Propulsion Laboratory and from there to the Chandra Operations Control Center in Cambridge, Massachusetts.

Chandra's solar panels are positioned incorrectly in Fig. 7.12 above, which was made to show the spacecraft's entire working surface. Correctly positioned, the solar panels always face the Sun. In contrast, Chandra's X-ray mirrors always face away from the Sun.

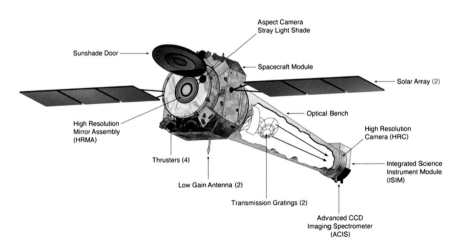

FIG. 7.12 Schematic diagram showing the various components of the Chandra X-Ray Observatory (Image credit: Harvard Smithsonian Center for Astrophysics)

Although Chandra was initially given an expected lifetime of 5 years, on 4 September 2001 NASA extended its lifetime to 10 years "based on the observatory's outstanding results." Physically Chandra could last much longer. A study performed at the Chandra X-ray Center indicated that the observatory could last at least 15 years. In July 2008, the International X-ray Observatory, a joint project between ESA, NASA and JAXA, was proposed as the next major X-ray observatory but was later cancelled. ESA later resurrected the project as the Advanced Telescope for High Energy Astrophysics (ATHENA+), with a proposed launch in 2028. At the time of writing, Chandra has exceeded everyone's expectations and the great X-Ray observatory is still going from strength to strength (Figs. 7.13 and 7.14).

One of the first pictures it took with its eight iridium coated mirrors was of Cassiopeia A, the remnants of a star that exploded in a supernova witnessed by Tycho Brahe in 1572. The picture was very beautiful, but more importantly, Chandra was already probing into Cassiopeia A's history. "Scientists can see evidence of what may be a neutron star or black hole near the center," NASA wrote in an August 1999 press release.

FIG. 7.13 The Shuttle astronaut crew that brought the Chandra X-Ray Observatory into orbit (Image credit: NASA)

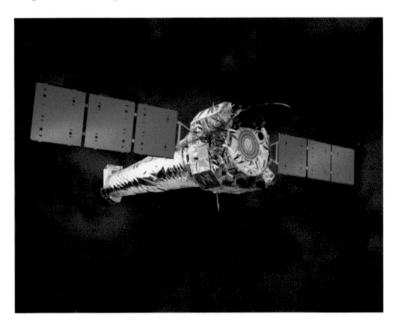

Fig. 7.14 Artists impression of the Chandra X-Ray Observatory in its highly elliptical orbit (Image credit: NASA)

Later that year, astronomers released a paper in *Astrophysical Journal Letters* discussing the chemical elements Chandra found in the gas surrounding the star. The findings included sulfur, silicon and iron that blasted out from the star's interior. Stars tend to burn off their hydrogen and helium earlier in their lifetimes; by the time these elements were fusing, temperatures in the star reached many billions of degrees before the explosion. Another of Chandra's early targets was the Crab Nebula, which showed—for the first time—a ring circling a pulsar star in the center of the nebula. Previously, Hubble spied wisps of matter surrounding the neutron, but the ring observed by Chandra was something entirely new.

"It should tell us a lot about how the energy from the pulsar gets into the nebula," said Jeff Hester, a professor at Arizona State University, in a September 1999 press release. "It's like finding the transmission lines between the power plant and the light bulb," he continued.

During its second year of operation, Chandra was going from strength to strength. Regular press conferences were arranged discussing the telescope's novel discoveries, including X-rays emanating from stars embedded in the Orion Nebula, a galaxy that was growing by snacking on one of its neighbors, and even evidence of new protostars.

Chandra soon began a series of discoveries concerning black holes. It unveiled evidence of a Type 2 quasar black hole emitting X-rays behind a thick sheet of material that up until then had hidden the black hole's existence. Later, scientists announced a possible new kind of black hole in the galaxy M82. From eight months of observations, the project scientists assigned to the telescope said the black hole could represent an evolutionary stage between small black holes formed from stars, and the much more massive ones lurking in the centers of galaxies.

"The black hole in M82 packs the mass of at least 500 suns into a region about the size of the moon," NASA reported in September 2000. Such a black hole would require extreme conditions for its creation, such as the collapse of a 'hyperstar' or the merger of scores of black holes.

Astronomers using Chandra also initiated a serious hunt for "dark" matter, which is believed to be practically invisible yet constitutes most of the universe. So far, we can only detect it through its gravity. In 2006, a team of astronomers spent more than 100 hours using Chandra to watch the galaxy cluster 1E0657-56, which contains gas from a galaxy cluster collision. Chandra's observations were combined with that of several other observatories.

Researchers examined the effect the galaxy cluster had on gravitational lensing, which is a known way that gravity distorts the light from background galaxies. Their observations of the gravitational field showed that normal matter and dark matter differentiated during the galaxy collision.

While the dark matter search continued, Chandra has been employed to find other missing matter. In 2010, researchers used Chandra and the European Space Agency's XMM-Newton observatory, probing a reservoir of gas resting along a wall of galaxies about 400 million light-years away from Earth.

Scientists found evidence of baryons—electrons, protons and other particles that compose matter found throughout much of our universe. The researchers suspected the gas would contain a significant amount of this matter.

While scientists continue to probe the nature of matter, Chandra continues to produce stunning pictures that also revealed the fabric of the universe. These pictures include a survey of planetary nebulae and a fast-growing galaxy cluster, as well as a "super-bubble" found in the Large Magellanic Cloud. Chandra's mission, originally expected to last five years and then extended to at least 10, is still going strong after more than 16 years in operation. The observatory has helped scientists glimpse the universe in action. It has watched galaxies collide, observed a black hole with cosmic hurricane winds, and glimpsed a supernova turning itself inside out after an explosion.

Astronomers have studied 51 quasars with NASA's Chandra X-ray Observatory and found they may represent an unusual population of black holes that consume excessive amounts of matter. As we have seen, quasars are objects that have supermassive black holes that also shine very brightly in different types of light. By examining the X-ray properties with Chandra, and combining them with data from ultraviolet and visible light observations, scientists are trying to determine exactly how these large black holes grow so quickly in the early universe.

The quasars that comprised this study—including the three shown as Chandra images in the bottom of the graphic—are located between about 5 billion and 11.5 billion light years from Earth. These quasars were selected because they had unusually weak emission from certain atoms, especially carbon, at ultraviolet wavelengths. Also, about 65 % of the quasars in this new study were found to be much fainter in X-rays, by about 40 times on average, than typical quasars.

The weak ultraviolet atomic emission and X-ray fluxes from these objects could be an important clue to the question of how a supermassive black hole accretes matter. Computer simulations show that, at least at low inflow rates, matter swirls toward the black hole in a thin disk. However, if the rate of inflow is high, the disk can puff up dramatically into a torus or donut that surrounds the inner part of the disk.

X-rays, produced in the region very near to the black hole are substantially blocked by the thick, donut-shaped part of the disk, making the quasar unusually faint in X-rays. This radiation is also prevented from striking the particles that are being blown away from the outer parts of the disk in a wind. This results in fainter ultraviolet emission from elements such as carbon.

The important implication of this work is that these "thick-disk" quasars may harbor black holes growing at an extraordinarily rapid rate. The current study, together with data collated from previous missions, suggested that such quasars might have been more common in the early universe, only about a billion years after the Big Bang. Such rapid growth might also explain the existence of huge black holes at even earlier times.

The destruction of a planet may sound like the stuff of science fiction, but a team of astronomers has found evidence that this may have happened in an ancient cluster of stars at the edge of the Milky Way galaxy. Using several telescopes, including NASA's Chandra X-ray Observatory, researchers have found evidence that a white dwarf star—the dense core of a star like the Sun that has run out of nuclear fuel—may have ripped apart a planet as it edged too close to its death star. But how could a white dwarf star, which is only about the size of the Earth, be responsible for such an extreme act? The answer, in a word, is gravity. When a star reaches its white dwarf stage, nearly all of the material from the star is packed inside a radius one hundredth that of the original star. This means that, for close encounters, the gravitational pull of the star and the associated tidal forces it generates, pulling on the near and far side of the planet, are greatly enhanced. For example, the gravity at the surface of a white dwarf is over ten thousand times higher than the gravity at the surface of the Sun.

Researchers used the European Space Agency's INTErnational Gamma-Ray Astrophysics Laboratory (INTEGRAL) to discover a new X-ray source near the center of the globular cluster NGC 6388. Optical observations had hinted that an intermediate-mass black hole with mass equal to several hundred solar masses or more resides at the center of NGC 6388. The X-ray detection by ESA's INTEGRAL then raised the intriguing possibility that the X-rays were produced by hot gas swirling towards an intermediate-mass black hole.

In a follow-up observation, Chandra's superlative X-ray vision enabled the astronomers to determine that the X-rays from NGC 6388 were not derived from the putative black hole at the center of the cluster, but instead from a location slightly off to one side.

FIG. 7.15 The Globular Cluster NGC 6388 contains a white dwarf that may have shredded a planet passing too near it. The image is a composite of visible and X-ray wavelengths (Image credit: NASA)

A new composite image showed NGC 6388 with X-rays detected by Chandra in pink, with visible light from the Hubble Space Telescope in red, green, and blue, with many of the stars appearing to be orange or white. Overlapping X-ray sources and stars near the center of the cluster also causes the image to appear white (Fig. 7.15).

With the central black hole ruled out as the potential X-ray source, the hunt continued for clues about the actual source in NGC 6388. It was monitored with the X-ray telescope on board NASA's Swift Gamma Ray Burst mission for about 200 days after the discovery by ESA's INTEGRAL. The source became dimmer during the period of Swift (discussed in an earlier chapter) observations. The rate at which the X-ray brightness attenuated was found to be in broad agreement with theoretical models of a disruption of a planet by the gravitational tidal forces of a white dwarf. In these models, a planet is first pulled away from its parent star by the gravity of the dense concentration of stars in a globular cluster. When such a planet passes too close to a white dwarf, it can be torn apart by the intense tidal forces of the white dwarf. The planetary debris is then heated and glows in X-rays as it falls onto the white dwarf. The observed amount of X-rays emitted at different energies agrees with expectations for a tidal disruption event. The researchers estimate that the destroyed planet would have contained about a third of the mass of Earth, while the white dwarf has about 1.4 times the Sun's mass.

While the case for the hypothetical tidal disruption of a planet was not quite iron-clad, evidence for this scheme of events was strengthened when astronomers used data from multiple telescopes to help eliminate other possible explanations for the detected X-rays. For example, the source did not reveal the distinctive features of a binary system containing a neutron star, such as pulsations or rapid X-ray bursts. Also, the source is much too faint in radio waves to be part of a binary system with a stellar-mass black hole.

Situated about 130 million light years from Earth, in the constellation of Canis Major, lies a pair of spiral galaxies—NGC 2207 and IC 2163—currently in the midst of a grazing encounter. In the course of just 15 years, NGC 2207 and IC 2163 have hosted no fewer than three supernova explosions and have produced one of the most bountiful collections of super bright X-ray sources known. These special objects—known as "ultra-luminous X-ray sources" (ULXs)—have been found using data from NASA's Chandra X-ray Observatory. As in our Milky Way galaxy, NGC 2207 and IC 2163 are peppered with many star systems known as X-ray binaries, which consist of a star in a tight orbit around either a neutron star or a "stellar-mass" black hole. The strong gravitational field of the neutron star or black hole pulls matter from the companion star. As this matter falls toward the neutron star or black hole, it is heated to millions of degrees, resulting in the emission of a torrent of X-rays.

Fig. 7.16 This composite image of NGC 2207 and IC 2163 contains Chandra data in *pink*, optical light data from the Hubble Space Telescope in *red*, *green*, and *blue* (appearing as *blue*, *white*, *orange*, and *brown*), and infrared data from the Spitzer Space Telescope in *red* (Image credit: NASA)

ULXs have far brighter X-rays than most 'normal' X-ray binaries. The true nature of ULXs is still debated, but they are likely to be a peculiar type of X-ray binary. The black holes in some ULXs may be heavier than stellar mass black holes and could represent a hypothesized, but as yet unconfirmed, intermediate-mass category of black holes (Fig. 7.16).

The Chandra composite image shown above contains about five times more observing time than previous efforts to study ULXs in this galaxy pair. Scientists now tally a total of 28 ULXs between NGC 2207 and IC 2163. Twelve of these vary over a period of several years, including seven that were not detected before because they were in a quiescent phase in earlier images.

The astronomers involved in studying these interacting galaxies had noted that there was a strong correlation between the number of X-ray sources in different regions and the rate at which stars were forming in same regions. The composite image shows this correlation, with the X-ray sources concentrated in the spiral arms of the galaxies, where large amounts of gas and dust is available for new rounds of star formation. This correlation also suggested that the companion stars in the binary systems were invariably young and massive.

Colliding galaxies like NGC 2207 and IC 2167 are well known to contain intense star formation. Shock waves—like the sonic booms from supersonic aircraft—are produced during the collision, leading to the collapse of clouds of gas and the formation of star clusters. In fact, researchers estimate that the stars associated with the ULXs are very young indeed, and may only be about 10 million years old. In contrast, our Sun is about halfway through its 10-billion-year lifetime. Moreover, analysis shows that stars of various masses are forming in this galaxy pair at a rate equivalent to form 24 stars the mass of our sun per year. In comparison, a galaxy like our Milky Way is expected to spawn new stars at a rate equivalent to only about one to three new suns per annum.

We are now ready to move to an entirely new region of the electromagnetic spectrum—microwaves, which are completely invisible to the human eye. Such a region has allowed astronomers and cosmologists to make great new strides in understanding the origin and nature of our universe, hearkening back to just a few thousand years after the Big Bang.

8. Probing the Microwave Sky

This author vividly remembers assisting his late father in tuning an old portable TV. When the TV was tuned in, we received good reception, but when it was tuned out, all we could make out was a constant hissing noise. Little did we know that what we were listening to was the cosmic background radiation, the 'afterglow of creation,' the microwave residual of the primordial fireball that was the hot Big Bang.

The cosmic background radiation was actually discovered serendipitously by the Canadian astrophysicist Andrew McKellar back in 1940. Observing cyanogen molecules among the interstellar clouds of our Milky Way, he derived their temperature. Intriguingly, this temperature was the same everywhere, that is, approximately 2.7 K (2.7 degrees above absolute zero). However, McKellar did not propose a mechanism causing the excitation of the molecules, but he correctly commented upon it as being the "temperature of space", maintained above absolute zero by a mysterious origin.

In 1948, the physicists Alpher, Bethe and Gamow published a milestone paper regarding the formation of elements in the newborn universe, during a time when it was very dense and very hot. They concluded that the Big Bang must have left behind a kind of 'fossil' heat, like the ashes of a fire extinguished long ago. The landmark paper predicted the existence of remnant radiation filling the universe, corresponding to a black body temperature of 5 K. Curiously, no one bothered to observe it at the time and no connection was made to McKellar's earlier work.

This radiation was re-discovered 17 years later by the American radio astronomers Arno Penzias and Robert Wilson, based at Bell Laboratories, Murray Hill, New Jersey, as uniform radio 'background noise' filling the entire sky. They won the Physics Nobel Prize in 1978 for their discovery and its interpretation (Fig. 8.1).

© Springer International Publishing Switzerland 2017
N. English, *Space Telescopes*, Astronomers' Universe,
DOI 10.1007/978-3-319-27814-8_8

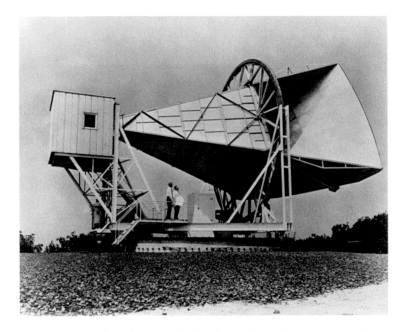

FIG. 8.1 Penzias and Wilson used this horn-shaped antenna to detect the cosmic microwave background radiation (Image credit: NASA)

This radiation arises from the time the universe was about 380,000 years old, when matter transitioned from opaque to transparent. One way to describe the Cosmic Microwave Background (CMB) is as the memorized printout of the last reflection of the primordial light of the early universe. This light existed when the electrons were still free and unbounded to the nuclei of atoms. Today, after over 13 billion years of expansion, this radiation has cooled enormously with a maximum intensity now at radio wavelengths, that is, in the centimeter and millimeter wavelength range. Today, the CMB glows at a maximum intensity at 7 cm radio lengths.

But there is much more to the CMB than meets the eye. The resolution of ground based telescopes was not sufficient to pick up any irregularities within this primordial radiation. This called for the construction of space based missions to study the phenomenon in unprecedented detail. The time was ripe for observational cosmology to become a precision science. Enter the Cosmic Background Explorer (COBE) satellite.

NASA accepted the proposal to build such a satellite, provided that the cost be kept under $30 million, excluding launcher and data analysis. Due to cost overruns in the Explorer program due to the Infrared Astronomical Satellite (discussed at length in a previous chapter), work on constructing the satellite at Goddard Space Flight Center (GSFC) did not begin until 1981.

The purpose of the Cosmic Background Explorer (COBE) mission was to take precise measurements of the diffuse radiation between 1 micrometer and 1 cm over the entire celestial sphere. The following quantities were measured: (1) the spectrum of the 3 K radiation over the range 100 micrometers to 1 cm; (2) the anisotropy (the property of being directionally dependent) of this radiation from 3 to 10 mm; and, (3) the spectrum and angular distribution of diffuse infrared background radiation at wavelengths from 1 to 300 micrometers.

The experiment module contained the instruments and a dewar filled with 650 liters of 1.6 K liquid helium, with a conical Sun shade. The base module contained the attitude control, communications and power systems. The satellite rotated slowly about the axis of symmetry to control systematic errors in the anisotropy (directionally dependent) measurements and to allow observations of the zodiacal light at various solar elongation angles. The orientation of the spin axis was maintained anti-Earth and at 94 degrees to the Sun-Earth line. The operational orbit was dawn-dusk Sun-synchronous so that the Sun was always to the side and thus was shielded from the instruments. With this orbit and spin-axis orientation, the instruments performed a complete scan of the celestial sphere every six months.

Originally, the COBE satellite was scheduled for launch by the Space Shuttle mission STS-82-B in 1988 from Vandenberg Air Force Base, but the Challenger explosion delayed this plan when the Shuttles were grounded. NASA kept their army of engineers from going to other space agencies to launch COBE, but eventually, a redesigned COBE was to be placed into a sun-synchronous orbit aboard a Delta rocket. COBE was launched November 18, 1989 and carried three instruments, a Far Infrared Absolute Spectrophotometer (FIRAS), to compare the spectrum of the cosmic microwave background radiation with a precise blackbody, a Differential Microwave

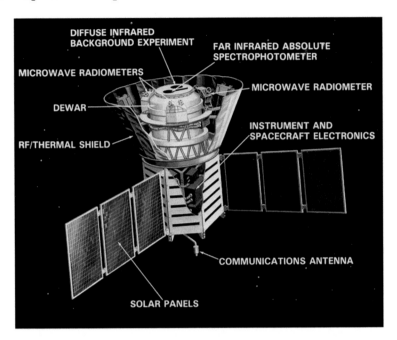

FIG. 8.2 Artist's schematic of NASA's COBE satellite (Image credit: NASA)

Radiometer (DMR) to map the cosmic radiation precisely, and a Diffuse Infrared Background Experiment (DIRBE) to search for the cosmic infrared background radiation (Fig. 8.2).

The need to control and measure all the sources of systematic errors required a rigorously integrated design. COBE would have to operate for a minimum of 6 months, carefully controlling the amount of radio interference from the ground, and from other satellites as well as radiation interference from the Earth, Sun and Moon. This required that the instruments on board the satellite be thermally stable as well as maintaining a high level of cleanliness to minimize entry of stray light and thermal emission from particulates.

In order to remove systematic errors in the measurement of the CMB anisotropy and measure the zodiacal cloud at different elongation angles for subsequent modeling required that the satellite rotate at a 0.8 rpm spin rate. The spin axis is also tilted back from the orbital velocity vector as a precaution against possible deposits of residual atmospheric gas on the optics as well against the infrared glow that would result from fast neutral particles hitting its surfaces at extremely high speed.

In order to meet the twin demands of slow rotation and three-axis attitude control, a sophisticated pair of yaw angular momentum wheels were employed with their axes oriented along the spin axis. These wheels were used to carry an angular momentum opposite that of the entire spacecraft in order to create a zero net angular momentum system.

The orbit would prove to be determined based on the specifics of the spacecraft's mission. The overriding considerations were the need for full sky coverage, the need to eliminate stray radiation from the instruments and the maintenance of thermal stability of both the dewar and the instruments. A circular, Sun-synchronous orbit would meet all these requirements. A 900 km altitude orbit with a 99° inclination was chosen as it fit within the capabilities of either a Shuttle (with an auxiliary propulsion on COBE) or a Delta rocket. This altitude was a good compromise between Earth's radiation and the charged particles in Earth's radiation belts at higher altitudes. An ascending node at 6 p.m. was chosen to allow COBE to follow the boundary between sunlight and darkness on Earth throughout the year.

The orbit combined with the spin axis made it possible to keep the Earth and the Sun continually below the plane of the shield, allowing an all-sky survey to be completed every six months. The last two important parts pertaining to the COBE mission were the dewar and Sun-Earth shield. The dewar, with its 650-liter superfluid helium cryostat, was designed to keep the FIRAS and DIRBE instruments cooled during the duration of the mission. It was based on the same design as one used on IRAS and was able to vent liquid helium along the spin axis near the communication arrays. The conical Sun-Earth shield protected the instruments from direct solar and Earth based radiation as well as radio interference from Earth and the COBE's transmitting antenna. Its multilayer insulating blankets provided thermal isolation for the dewar.

The science mission was conducted by the three instruments detailed previously: DIRBE, FIRAS and the DMR. The instruments overlapped in wavelength coverage, providing a consistency check on measurements in the regions of spectral overlap and assistance in discriminating signals from our galaxy, Solar System and CMB.

After its launch in November of 1989, COBE set to work to test the 'standard model' of cosmology. In simple terms, the model says that our universe was born some 15 billion years ago in a hot Big Bang. This explosion from an unimaginably hot, dense state created not only matter and radiation but also space and time. In the eons that have elapsed since these early times, the universe has been expanding and cooling. Central to this model of cosmology was the nature and structure of the CMB.

Only two months into its mission, COBE showed that the background follows the spectrum of a 2.73 K blackbody to high accuracy. According to the cosmologist P.J.E. Peebles of Princeton University, "this evidence that the cosmos was once was hot and dense is as tangible as the dinosaur footprints that show how different the inhabitants of the Earth were 100 million years ago."

On large scales, the universe consists of galaxy clusters arranged in enormous bubbles, walls and voids. If it started out as a uniform soup of matter and radiation, as hot Big Bang cosmology predicts, how did these structures arise? According to the standard model, they are supposed to have grown from miniscule density irregularities in the primeval fireball.

Regions that happened to end up with slightly more matter than average acted as 'seeds' in that their extra gravity attracted still more matter. These primeval density fluctuations should have imprinted themselves on the background radiation as minute variations in temperature. COBE's differential microwave radiometers, or DMRs, were designed to detect these fluctuations. Maps of the microwave sky made from the first year of DMR's data looked perfectly smooth at first. Then Smoot and his colleagues set about searching for the proverbial needle in a haystack. So they began subtracting the uniform brightness of a 2.73 K blackbody from their maps. Next they compensated for our motion through space, which imparts a small Doppler shift to the data and causes the sky to appear to be a bit warmer in the direction we're travelling in and cooler behind. Their data reduction techniques also had to allow for every potential systematic error they could think of, including the effects of solar activity and interference from powerful radars on the Earth. Finally they modeled all the known microwave emissions of the Milky Way and subtracted them away too (Fig. 8.3).

The resulting map (shown above) revealed what remained for the observations made at a wavelength of 5.7 millimeters. The sky appeared mottled with temperature variations measured in tens of

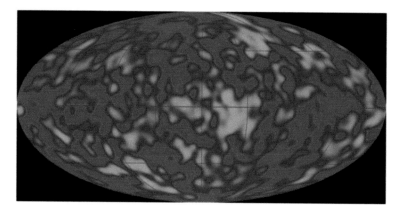

FIG. 8.3 The famous all-sky microwave map captured by the COBE satellite. The colors depict minute temperature variations (depicted here as varying shades of *blue* and *purple*) are linked to slight density variations in the early universe. These variations are believed to have given rise to the structures that populate the universe today: clusters of galaxies, as well as vast, empty regions (Image credit: BBC)

millionths of a degree. By comparing the signals from different DMR antennae and receivers, Smoot and his co-workers estimated that most of the fluctuations were instrument noise. The rest, they claimed, were the long sought after cosmic density fluctuations with a typical deviation of only 0.00003 K.

The mottling in the COBE map occurs in patches some 10 to 90 degrees across. When the radiation was emitted, the corresponding regions of space were too large for light to have crossed since the Big Bang. Thus it was hard to see how the cosmic background could have precisely the same temperatures over such large angles.

Unbeknown to them, this problem was actually anticipated and solved more than a decade before, when Alan H. Guth (MIT) proposed that our present universe 'inflated' from a tiny volume of space-time some 10^{-35} to 10^{-30} seconds after the Big Bang. In this scenario the temperature variations observed by COBE were the enlarged relics of microscopic quantum fluctuations present at the end of the inflationary era. With inflation taken into account, the standard model predicted that the primeval density fluctuations would be equally strong on all angular scales. The DMR data seem to fit this 'scale invariance.' But several alternative theories of the origin and structure also predicted such behavior. Thus, COBE's discovery didn't prove inflation, but was consistent with it.

The project's findings lent further credence to the popular notion that most of the matter in the universe exists in some unknown and invisible form. This is because such subtle density enhancements could not have grown into the largest observed concentration of galaxies in the time since the Big Bang. If the seeds were actually much denser, they must have formed mostly of a kind of matter that doesn't interact with radiation as strongly as ordinary matter. In other words, what the COBE map demonstrated was that gravity can make the structure we see if dark matter were indeed present. The precise nature of this mysterious matter remains one of the biggest unresolved mysteries in cosmology.

Other cosmologists compared the new COBE measurements with the predictions of nearly 50 different variations of the standard model. Some models entertained 'cold' dark matter, made of hypothetical weakly interacting particles, while others modeled 'hot' dark matter consisting of massive neutrinos. To date, no class of theory has been ruled out categorically. But what is widely agreed among cosmologists is that a combination of the Big Bang, inflation, and dark matter as the simplest and best explanation of what the COBE all sky map depicts.

In the years leading up to the launch of COBE, there were two significant astronomical developments. First, in 1981, two teams of astronomers, one led by David Wilkinson of Princeton and the other by Francesco Melchiorri of the University of Florence, simultaneously announced that they detected a quadrupole distribution of CMB using balloon-borne instruments. These findings foreshadowed the detection of the black-body distribution of CMB that FIRAS on COBE was to measure. In particular, the Florence group claimed a detection of intermediate angular scale anisotropies at the level 100 micro Kelvins, in agreement with later measurements made by the BOOMERanG (discussed later in this chapter) experiment.

However, a number of other experiments attempted to duplicate their results and were unable to do so. What's more, in 1987 a Japanese-American team led by Andrew Lange and Paul Richards of UC Berkeley and Toshio Matsumoto of Nagoya University published data from sounding rocket experiments that showed an excess brightness at 0.5 and 0.7 mm wavelengths. Their measurements implied that the CMB was not that of a true black body.

With these developments serving as a backdrop to COBE's mission, scientists eagerly awaited results from FIRAS which involved comparing raw data from a 7° patch of the sky against an internal black body "control". The interferometer in FIRAS covered between 2 and 95 per cm in two bands separated at 20 per cm. The data was collated continuously over a ten-month period and it showed a perfect fit of the CMB and the theoretical curve for a black body at a temperature of 2.7 K, thus casting doubt on the Berkeley-Nagoya results erroneous.

The DMR devoted a full four years to mapping the detectable anisotropy of cosmic background radiation as it was the only instrument not dependent on the dewar's supply of helium to keep it cooled. This operation was able to create full sky maps of the CMB by subtracting out galactic emissions at various frequencies. It's important to remember that the cosmic microwave background fluctuations are extremely faint, only one part in 100,000 compared to the 2.73 K average temperature of the radiation field. The density ripples are believed to have produced structure formation on the largest scales observed in the universe, including clusters of galaxies and so-called voids—vast regions utterly devoid of galaxies.

DIRBE also detected 10 new far-infrared emitting galaxies in the region not surveyed by IRAS (discussed earlier in the book) as well as nine other candidates in the weak far-infrared that may be spiral galaxies. Galaxies that were detected at the 140 and 240 micron range were also able to provide information on very cold dust. At these wavelengths, the mass and temperature of this dust can be derived. When these results were collated with 60 and 100 μm data taken from IRAS, it was found that the far-infrared luminosity arises from cold ($\approx$17–22 K) dust associated with diffuse HI cirrus clouds, 15–30 % from cold ($\approx$19 K) dust associated with molecular gas, and less than 10 % from warm ($\approx$29 K) dust in the extended low-density HII regions.

As well as providing new insights on distant galaxies, DIRBE also made two other significant contributions to our understanding of the Solar System. For example, it was able to conduct studies on interplanetary dust (IPD) and determine if its origin were derived from asteroid or particles of comets. The DIRBE data collected at

12, 25, 50 and 100 micron surveys and these data pointed to an asteroidal origin for this interplanetary dust.

The second contribution DIRBE made was to model the Galactic disk as seen edge-on from our position, some 26,000 light years from its center. According to the model, if our Sun is 8.6 kiloparsecs (one parsec is 3.26 light years) from the Galactic center, then the Sun would have to be 15.6 parsecs above the mid-plane of the disk, which has radial and vertical scale lengths of 2.64 and 0.333 kiloparsecs, respectively, and is warped in a way consistent with the HI layer. Curiously, DIRBE provided no indication of a thick disk.

To create this model, the IPD had to be subtracted out of the DIRBE data. It was found that this cloud, which as is more familiarly called the Zodiacal light, was not centered on the Sun, as previously thought, but on a locus in space a few million kilometers away. Astronomers attribute this offsetting to the gravitational influence of Saturn and Jupiter.

Despite making considerable progress in understanding the origin and nature of the CMB, there were numerous cosmological questions left unanswered by COBE's results. A direct measurement of the extragalactic background light (EBL) could also provide important constraints on the cosmological history of star formation, metal and dust production, as well as the conversion of starlight at visible wavelengths into infrared emissions by dust.

By looking at the results from DIRBE and FIRAS in the 140 to 5000 micron range, astronomers were able to estimate the extent to which primordial matter had been processed inside stars. The data indicated that about 20–50 % of the total energy released in the formation of helium and metals throughout the history of the universe. Deriving only from nuclear sources, this intensity implies that more than 5–15 % of the baryonic mass density produced by big bang nucleosynthesis had been processed in stars to create helium and heavier elements.

COBE also made significant advances in our understanding of star formation. It provided important constraints on the cosmic star formation rate. Observations made by COBE required that the star formation rate at redshifts of $z \approx 1.5$ to be larger than that inferred from UV-optical observations by a factor of two. This excess stellar energy must be mainly generated by massive stars cocooned inside as yet undetected dust enshrouded galaxies or extremely dusty star forming regions in observed galaxies. The exact star formation history could not be unambiguously pinned down by COBE however, and further observations must be made in the future.

Needless to say, the COBE mission was a spectacular success. A team of American scientists announced, on April 23, 1992, that they had found the primordial "seeds" (CMBE anisotropy) in data from COBE. The announcement was reported worldwide as a fundamental scientific discovery and ran on the front page of the *New York Times*. Instrument operations were terminated December 23, 1993. As of January 1994, engineering operations were to cease later that month, after which operation of the spacecraft was transferred to Wallops for use as a test satellite.

On June 30, 2001, NASA launched a follow-up mission to COBE led by DMR Deputy Principal Investigator Charles L. Bennett. The Wilkinson Microwave Anisotropy Probe (WMAP), launched on June 30, 2001, has clarified and expanded upon COBE's accomplishments. The spacecraft was positioned near the second Lagrangian point (L2), a gravitational balance point between Earth and the Sun and 1.5 million km (0.9 million miles) opposite the Sun from Earth. This orbit helped isolate WMAP from radio emissions from Earth and the Moon without having to place it on a more distant trajectory that would complicate tracking. WMAP was initially planned to operate for two years, but its mission was extended to Sept. 8, 2010. After its mission ended, WMAP moved from L2 into a standard orbit around the Sun (Fig. 8.4).

The spacecraft carried a pair of microwave receivers that observed in nearly opposite directions through 1.4×1.6-metre (4.6×5.2-foot) reflecting telescopes. These reflectors operated much like a commercial satellite "dish" antenna and measured the relative brightness of opposite points in the universe at frequencies of 23, 33, 41, 61, and 94 gigahertz and were cooled to eliminate internal noise. The spacecraft was protected from the Sun by a shield that was deployed with the solar arrays and was permanently pointed at the Sun. The spacecraft rotated so the two reflectors scanned a circular patch of sky. As WMAP orbited the Sun near the L2 point, the scanned circle was allowed to precess in such a way that the entire sky was mapped every six months. Occasionally, the planet Jupiter would pass through the field of view, allowing it to act as a calibration source (Fig. 8.5).

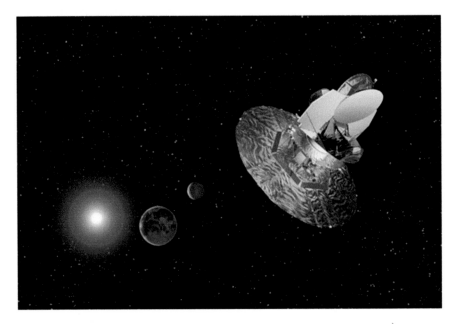

Fɪɢ. **8.4** Artist's impression of WMAP in orbit (Image credit: NASA)

Fɪɢ. **8.5** A sky map constructed from data collated by WMAP. The image reveals 13.77 billion year old temperature fluctuations (shown as color differences) that correspond to the seeds that grew to become the galaxies. The signal from our Galaxy was subtracted using the multi-frequency data. This image shows a temperature range of ±200 micro Kelvin (Image credit: NASA)

WMAP measured small variations in the temperature of the cosmic microwave background radiation. These variations were minute: one part of the sky has a temperature of 2.7251 K (degrees above absolute zero), while another part of the sky has a temperature of 2.7249 K. In contrast to its COBE its predecessor, WMAP measured anisotropy with much finer detail and greater sensitivity.

WMAP's map of the neonatal universe revealed the afterglow of the hot, young universe at a time when it was only 375,000 years old, when it was a tiny fraction of its current age of 13.77 billion years. WMAP data placed severe constraints on the cosmic events that could have possibly happened earlier, and what could have happened in the billions of year since that early epoch. WMAP lent greater credence to the Big Bang paradigm, which posits that the young universe was initially very hot and dense, and which has been expanding and cooling ever since. In addition observations carried out by WMAP provided new insights into the size, matter content, age, geometry and fate of the universe. Specifically, only 4.6 % of the matter that makes up the observable cosmos consists of ordinary atoms. A much greater fraction—24 % of the universe—is a different kind of matter that exerts a gravitational influence but does not emit any light. This is the elusive called "dark matter". However, by far the largest fraction (71 %) of the current composition of the universe is a mysterious kind of energy with anti-gravity properties that is driving the acceleration of the expansion of the universe. In their ignorance, cosmologists call this component dark.

Following WMAP, the European Space Agency's probe, Planck has continued to increase the resolution at which the background has been mapped. Launched in 2009 Planck mapped the anisotropies of the CMB at microwave and infra-red frequencies, with high sensitivity and small angular resolution. The mission substantially improved upon observations made by WMAP. Planck provided a major source of information relevant to several cosmological and astrophysical issues, such as testing theories of the early universe and the origin of cosmic structure; and by 2013 it succeeded in providing the most accurate measurements of several key cosmological parameters, including the average density of ordinary matter and dark matter in the universe.

Following fast on the heels of WMAP, the Planck project was initiated around 1996 and was initially called COBRAS/SAMBA: the Cosmic Background Radiation Anisotropy Satellite/Satellite for Measurement of Background Anisotropies. It was later renamed in honor of the German physicist Max Planck (1858–1947), who, as we have seen in Chap. 1, derived the formula for black-body radiation.

Built at the Cannes Mandelieu Space Center by Thales Alenia Space, and created as a medium-sized mission for ESA's Horizon 2000 long-term scientific program, Planck was launched in May 2009, reaching the Earth/Sun L2 point by July, and by February 2010 had successfully started a second all-sky survey. On 21 March 2013, the mission's first all-sky map of the cosmic microwave background was released, with an expanded release including polarization data in February 2015. All the data is anticipated to be analyzed by 2016.

When its mission was terminated, Planck was redirected into a heliocentric orbit and passivated in order to prevent it from endangering any future missions. The final deactivation command was sent to Planck in October 2013 (Fig. 8.6).

Fig. 8.6 Artist's impression of the Planck satellite at the L point above Earth (Image credit: ESA)

The spacecraft carried two instruments: the Low Frequency Instrument (LFI) and the High Frequency Instrument (HFI). Both instruments could detect both the total intensity and polarization of photons, and together covered a frequency range of nearly 830 GHz (from 30 to 857 GHz). The CMB spectrum peaks at a frequency of 160.2 GHz.

Planck's passive and active cooling systems enabled its instruments to maintain a temperature of −273.05 °C (−459.49 °F), or just 0.1 degrees Celsius above absolute zero. This meant that after August 2009, the Planck satellite was the coldest known object in space, until its active coolant supply was exhausted in January 2012.

Although primarily a European mission, NASA also played a role in the development of Planck by putting their weight behind the analysis of the scientific data it collected. The Jet Propulsion Laboratory built components for the Planck science instruments, including bolometers for the high-frequency instrument, a 20 K cryo-cooler for both the low- and high-frequency instruments, and an amplifier technology for the low-frequency instrument.

The mission had a wide variety of scientific aims, including: high resolution detections of both the total intensity and polarization of primordial CMB anisotropies, the creation of a catalogue of galaxy clusters through the Sunyaev–Zel'dovich, caused by high energy electrons distorting the cosmic microwave background radiation. Planck also carried out observations of the gravitational lensing of the CMB, as well as monitoring the bright extragalactic radio (active galactic nuclei) and infrared (dusty galaxy) sources. It made important observations of the Milky Way's magnetic field, the interstellar medium, and the distribution of synchrotron emissions in the Galaxy. Closer to home, Planck carried out studies of the Solar System, including planets, asteroids, comets and the zodiacal light.

The satellite had a higher resolution and sensitivity than WMAP, allowing it to probe the power spectrum of the CMB to scales three times smaller than the former. It also observed in nine frequency bands compared to WMAP's five. It was widely anticipated that most Planck measurements would be limited by how well foreground signals could be subtracted out, rather than by the detector performance or length of the mission. This was particularly important for the polarization measurements. The predominant foreground radiation depends on frequency, but could include synchrotron radiation from the Milky Way at low frequencies, and dust at high frequencies.

Planck started its first all-sky survey on 13 August 2009. By September of the same year, ESA announced the preliminary results from the Planck First Light Survey, which was performed to demonstrate the stability of the instruments and the ability to calibrate them over long periods. Early testing suggested all was well and that the data quality would be excellent.

On 15 January 2010, the mission was extended for another year, with observations continuing until the end of 2011. After the successful conclusion of the first survey, the spacecraft started a second all sky survey on 14 February 2010, having already surveyed 95 % of the sky and 100 % sky coverage completed by mid-June of 2010.

On 17 March 2010, the first Planck images were made available to the public, showing dust concentration within 500 light years from the Sun. On 5 July of the same year, the Planck team delivered its first all-sky image. Many more exciting images from the satellite were presented to astronomers attending the January 2011 Planck conference in Paris. And on May 5 2014 a Planck-derived map of the galaxy's magnetic field was published.

On February 5 2015, a new map generated by the Planck satellite uncovered the polarized light from the early universe as seen across the entire sky, revealing that the first stars formed much later than previously thought. Ordinary light vibrates in every conceivable orientation, but polarized light vibrates in a preferred direction. This can arise as a result of photons—the 'particles' of light—bounce off other particles. This is exactly what happened in the early universe, when the CMB was created. Initially, photons were trapped in a hot, dense soup of particles that, by the time the universe was a few seconds old, consisted mainly of electrons, protons and neutrinos. Owing to the high density of matter in that early epoch, electrons and photons collided with one another so frequently that light could not travel any significant distance before bumping into another electron. As a result, the early universe could be said to be extremely 'foggy'. Slowly but surely though, as the cosmos expanded and cooled, photons and the other particles grew further apart, causing these collisions to occur less frequently. This had two consequences.

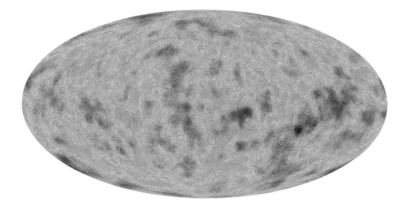

FIG. 8.7 A visualization of the polarization of the Cosmic Microwave Background, or CMB, as detected by ESA's Planck satellite over the entire sky. The color scale represents temperature differences in the CMB, while the texture indicates the direction of the polarized light. The patterns seen in the texture are characteristic of 'E-mode' polarization, which is the dominant type for the CMB (Image credit: ESA)

The first was that electrons and protons could finally combine and form neutral atoms without them being torn apart again by an incoming photon. In addition, photons had enough room to travel, being no longer trapped in the cosmic fog (Fig. 8.7).

The spectacular success of the Planck satellite helped bolster hot Big Bang cosmology, affirming that it is the single most powerful model of the origin and evolution of the universe we have at our disposal. The age of cosmos was pinned down to 13.799 ± 0.038 billion years. The rate of expansion was calculated to be 67.8 ± 0.9 km/s/megaparsecs, a quantity known to cosmologists as the Hubble parameter. Finally, Planck data helped provide a more accurate estimate of the contribution from the elusive dark energy to be $69.2 \pm 1.2\%$.

In the next chapter, we return to our own Solar system once more, and uncover how astronomers employed ever more sophisticated space based observatories to reveal fascinating secrets about our Sun, and by implication, other stars that inhabit the universe.

9. Empire of the Sun

The Sun, the star that brings life to our plant, is a sphere of mainly hydrogen and helium gas, generating heat by thermonuclear fusion, and rotating about once every 25 days at its equator, slowing to once every 34 days near its poles. The sunspots astronomers observe on the sun's surface (the photosphere) consist of a dark central area, called the umbra, surrounded by a lighter penumbral region. Sunspot numbers vary over a cycle of about 11 years, although this cycle has been shown to vary considerably from as long as 17 years to as short as 7 years. The cycle starts with spots appearing at about 30 to 40 degrees latitude, north and south of the Sun's equator, immediately after sunspot minimum. The latitude gradually reduces as the cycle progresses, decreasing to about 15 degrees at sunspot maximum, and just 6 degrees at sunspot minimum. Curiously, very few sunspots have been recorded at the solar equator. After minimum, sunspots then start to re-appear at mid-latitudes, after which time, the cycle continues apace.

Sunspots are a few thousand degrees cooler than the surrounding surface of the Sun, but they are also sites of intense magnetic activity being about 3000 gauss, compared with the one gauss of the rest of the Sun's photosphere. The spots tend to appear in pairs, with the preceeding, or 'p' spot of the pair, as it moves across the solar disk, almost always having a different polarity to the other spot. The p spots in one hemisphere of the Sun generally have the same polarity as each other, which is opposite to the p spots in the other hemisphere. When a new sunspot cycle starts, the polarities reverse between the hemispheres, so the sunspot cycle is really 22 years rather than an 11 year cycle, as is more commonly quoted.

Large loops and filaments rapidly eject gas away from the surface, manifesting themselves in the form of prominences at the edge of the Sun during total solar eclipses or using modern energy rejection filters coupled to telescopes. The most spectacular prominences, which can persist for anything from a few hours to a few

© Springer International Publishing Switzerland 2017
N. English, *Space Telescopes*, Astronomers' Universe,
DOI 10.1007/978-3-319-27814-8_9

days, are often associated with sunspots. Furthermore, they sometimes produce bright solar flares which typically last for a few minutes and which are best seen at one particular red wavelength of light known as hydrogen alpha (656.28 nm). In addition, many cloud-like prominences associated with these faculae are found on the surface of the Sun. These eruptive prominences contain hydrogen, helium, and minor amounts of other elements, including iron, titanium, calcium and barium. In contrast, the cloud-like prominences were shown to consist almost entirely of hydrogen, helium, with a trace of calcium thrown in for good measure. Sunspots and eruptive prominences are generally limited to low and mid-latitudes, whereas faculae and cloud-like prominences can be seen at any latitude.

Under normal circumstances, the very bright photosphere overwhelms the light from the solar chromosphere and corona, both of which can be observed during a total solar eclipse. The chromosphere is a thin, pale pink colored band surrounding the Sun, located just above the photosphere, while the corona consists of an extensive white halo that completely surrounds the Sun and extends out to several solar radii into the vacuum of interplanetary space. On closer inspection, the outer edge of the chromosphere is seen to consist of many fine, jet-like projections called spicules. The observed morphology of the corona changes as the solar cycle progresses from being roughly circular around solar maximum to decidedly elliptical near solar minimum.

The chromosphere can be imaged across the total solar disk, without waiting for a total solar eclipse, by taking photographs in hydrogen alpha light or at the wavelengths of the calcium H and K lines. These so-called spectroheliograms reveal long, arc-shaped filaments mentioned, together with bright plages which are associated with sunspots.

A correlation between sunspots and disturbances in the Earth's magnetic field was established as early as 1896, when the Norwegians Kristian Birkeland and Carl Stormer showed that cathode rays (essentially electrons) were emitted by the Sun. This work was followed up in 1904, when the British astronomer, E. Walter Maunder, garnered evidence that large magnetic storms on Earth initiate within thirty hours following the arrival of a large sunspot group at the center of the Sun's disk, suggesting that sunspots emitted some sort of particle that travels through interplanetary space to reach Earth. That said, the smaller, more frequent storms did not seem to be generally associated with sunspots, however, but they had a tendency to recur every 27 days, unlike the behavior of larger storms, which was correlated with the synodic period of rotation of the Sun's equator. The German geophysicist Julius Bartels called these invisible sources on the Sun 'M regions', on account of their strong association with magnetically active regions on the Sun.

As radio technology developed apace during the twentieth century, Howard Dellinger of the American Bureau of Standards showed as early as the 1930s that problems with the reception of short-wave radio signals on Earth were often, but not always, associated with solar flares seen on the Sun appearing a short time before. Subsequent work revealed that these radio waves were being reflected from the F region of the Earth's ionosphere. From these observations Dellinger was the first to suggest that some unknown solar phenomena was producing solar flares and emitting some form of invisible radiation that penetrated the F region, altering the structure of lower ionosphere in such a way that it caused it to become opaque to radio waves. Then in 1939, T.H. Johnson, working for the Bartol Research Foundation and Serge Korff of New York University, suggested that the origin of these disturbances was solar X-rays.

Progress in astronomy, like many other sciences, often comes as a result of serendipitous rather than deliberate observations. One such fortuitous discovery came in February 1942, when British army radar surveys picked up radio noise which was initially attributed to local jamming. James S. Hey, a physicist at the British Army's Operational Research Group, was assigned to investigate the cause of this interference and found that the source was the Sun. What is more, he found that this radio noise was correlated with a large sunspot group crossing the solar meridian. Further work in the following years established that solar flares emitted radio waves but, unlike the case of the sunspot-generated radio signals, these flares could pop up anywhere on the Sun's visible disk and still produce radio signals that could be picked up on Earth. Curiously, the radio signals from flares lasted only a few minutes, in contrast to those associated with sunspots, which could persist for hours. Thus, it was becoming clear that the Sun could not only interrupt short-wave communications on Earth by modifying the Earth's ionosphere, but it also produced radio noise of its own from time to time.

Further work conducted by the American scientist, Scott Ellsworth Forbush, in the 1940s discovered that the flux of medium and high energy cosmic rays observed at the Earth's surface varied according to the 11 year solar cycle, falling off slightly at the approach of solar maximum. In addition, these cosmic ray counts would often fall off suddenly during solar storms. Solar particles or X-rays were unlikely in themselves to have much of an effect, but the Sun's magnetic field would, by deflecting these cosmic rays away from the inner solar system. It was the Swedish electrical engineer, Hannes Alfven, who first suggested that the Sun's magnetic field was increased during active periods sending a shower of charged particles away from its surface and which would sometimes intercept the Earth.

In the early 1950s, the American father and son astronomers, Harold and Horace Babcock, of the Mount Wilson Observatory, took measurements of the Sun's magnetic field across its disk with their newly developed solar magnetograph, and were able to demonstrate that there were both bipolar and unipolar regions present, corresponding to magnetic fluxes which were either leaving or entering the Sun and which did not appear to be connected with sunspots at all. In the bipolar regions the magnetic flux leaving the Sun was about equal to that entering it.

The Babcocks suggested that ions and electrons (called corpuscular radiation) coming from the Sun in bipolar regions would follow a course along the field lines immediately above those regions where they collide with each other above the solar surface, generating radio noise as well as forming prominences and flares. On the other hand, the corpuscular radiation leaving the Sun in unipolar regions, would stream away from it and eventually reach Earth, modifying the ionosphere and causing problems with short-wave reception. The Babcocks also believed that these unipolar regions were Bartel's M regions. Corpuscular radiation would also leave the Sun from its polar regions and follow the general magnetic field lines far out into interplanetary space.

Thus far we have discussed energetic attributes of the Sun, such as sunspots, prominences, flares and their effect on our life-bearing world. But even when the Sun is in its quiescent phase, it still exerts its influence on our planet that goes well beyond its fundamental role in maintaining the Earth in its orbit and providing lie giving light and heat.

In June 1942 George C. Southworth of Bell Labs, discovered that the Sun emitted radio waves continuously in the centimeter wavelength band. In addition to this, scientists based in Australia and elsewhere also detected solar emissions in the meter band. These centimeter radio emissions were independently shown in 1946 by the Australian physicist, David F. Martyn, and the Russian astronomers, Vitaly Ginzburg and Iosif Shklovskii, to be derived from a region of the solar chromosphere, where temperature can reach 10,000 K, while the meter wave band emissions were believed to be produced by the still hotter corona at a temperature of some 2 million K.

The Sun is not only a continuous source of visible light and radio waves. This was made clear in 1952 when sounding rocket experiments showed that the solar corona also emits X-rays that are intense enough to sustain the E region—located between 90 and 160 kilometers altitude—of the Earth's ionosphere. In similar fashion, the Sun's far ultraviolet radiation, which does not penetrate our atmosphere, was found to sustain the D region of the ionosphere.

Back in the early seventeenth century, the German astronomer and mathematician, Johannes Kepler, suggested that cometary tails point away from the Sun because material coming from the head of the comet is blown outwards by particles of solar radiation. Kepler's ideas fell out of favor in the nineteenth century, however, as light was then thought to be a wave phenomenon and not a stream of particles. But his fecund idea enjoyed a new lease of life in the early twentieth century, however, when light was shown to exhibit both wave and particle-like characteristics, exerting pressure on objects in its path.

As the twentieth century unraveled, however, it became increasingly clear that sunspots probably emitted charged particles. What's more, some astronomers believed that there may be a weak flux of charged particles evaporating continuously from the Sun.

Some of these particles would be captured by the Earth's magnetic field, which would then channel them to the Earth's poles, causing the aurorae. But could the pressure of light itself really cause the tails of comets to point away from the Sun?

In 1951, Ludwig Biermann of the Max Planck Institute in Germany demonstrated that the so-called radiation pressure suggested by Kepler could not in fact exert enough force to push the tails of comets away from the Sun, but solar ions and electrons could if they could reach velocities between 500 and 1000 kilometers per second and only if their density was between 100 and 1000 per cubic centimeter at the distances of Earth's orbit. This would mean that our star would have to generate an average ion (or plasma) flux of about 2×10^{10} ions per square centimeter per second.

Intrigued by Edlen's very not solar corona of some 2 million K, the British geophysicist, Sydney Chapman, performed independent calculations which showed that a corona consisting mainly of electrons and protons, would indeed extend far beyond the Earth's orbit, where its temperature would still be very high (200,000 K.) Bierman's theory was based upon a solar environment that was essentially dynamic, emitting a continuous stream of charged particles. This was in sharp contrast to Chapman's theory where the corona was in a state of hydrostatic equilibrium, with the gravity of the Sun balancing the corona's gas pressure.

Intriguingly, while Biermann's and Chapman's assumptions were unquestionably different, their predicted plasma (or ion) densities at the distance from the Earth's orbit from the Sun were, to all intents and purposes, the same. That said, it soon became clear that Chapman's theory hit a snag, in that it predicted an ion pressure at the orbit of Pluto some 10 million times greater than that thought to exist in the interstellar medium! Needless to say, Chapman's predictions seemed unlikely, although it could not, as yet, be dismissed out of hand.

It was left to astrophysicist Eugene Parker, then (1957) based at the Enrico Fermi Institute, to reconcile these two apparently contradictory theories by adding dynamic terms to Chapman's static equations in order to better represent the hypothesized continuous expansion an evaporation of the solar corona. Parker found that his results fitted Biermann's predictions more than Chapman's. In 1958, Parker further refined his theory of the solar wind, showing that the expanding coronal gas should draw the magnetic field lines in the corona far out into the solar system, producing spiral magnetic field lines created by a rotating Sun. But Parker's theory was just too efficient at transferring energy through the corona, even using Biermann's lowest estimate of solar flux, so a highly efficient mechanism was required to maintain the Sun's corona at the 2 million K near the Sun, as estimated by Edlen. Parker then envisaged hydromagnetic waves which could move though the Sun's photosphere to supply the required energy. That said, his theory was rejected by the majority of physicists.

Thus, as the space age dawned upon the world, scientists had a good understanding of a great deal of solar phenomena, including sunspot variabilities, solar flares, prominences, Sun-derived radio noise, aurora storms near the Earth's magnetic poles and shortwave radio fades, but the relationship between these was only imperfectly understood. What's more, the origin of the sunspot cycle, as well as the cause of the ultra-high temperature of the corona still remained a mystery. Even the nature and character of the suspected solar wind was also unresolved. Only by studying these solar phenomena from the pristine environment of space could further advances be made in this arena of human enquiry.

As we have seen, the first spacecraft to leave the Earth's atmosphere were necessarily low Earth-orbiting, with the result that the *in situ* environment they were placed in was still under the influence of the Earth's protective magnetic field. The first spacecraft to measure the solar plasma stream with the planet's magnetic field was Luna 2 in September 1959, during its trip to the Moon. Its valuable data suggested that the extended solar corona, at the distance from the Earth to the Sun, did indeed consist of high speed ions, thereby confirming Biermann's theory. The flux measured was about 2×10^8 ions per square centimeter per second, however, or about two orders of magnitude lower than what Biermann had predicted. Alas, Luna 2 was not able to ascertain the direction of the ions in the solar wind directly equipped as it was with rather primitive ion traps. Further analysis of Luna 2's results by Shklovskii suggested that the primary reason for the low flux was due to a plasma density that was considerably lower than that predicted by Biermann.

In 1958 the Space Science Board of the National Academy of Sciences started considering in areas of priority research that should be undertaken in space. One suggestion that came from its Committee on Space Projects, chaired by Bruno Rossi, was a study of the interplanetary plasma, or solar wind but. That said, although this suggestion was forwarded by the Space Science Board, it was not deemed important enough for the newly formed NASA to endorse. The reason for their lukewarm attitude toward such a study was practically based. For one thing, the instrumentation would need developing, and it was no mean task at that time to deploy satellites sufficiently far outside the Earth's magnetic field. But NASA's initial recalcitrance did not last long, however. Political pressure in the late 1950s to expand the American space program, following Russia's series of dramatic space successes, meant that such a mission could be sanctioned after all. So NASA agreed to fund a team led by Herbert Bridge team at MIT, to design and build a plasma probe to hitch a ride on Explorer 10, after which it would be placed into a highly eccentric orbit around the Earth with an apogee of 180,000 kilometers (Fig. 9.1).

Bridge's plasma probe was a quantum leap forward on the instrument carried on the by Luna 2 spacecraft, with a much higher signal to noise ratio and capable of measuring the ion flux in six

FIG. 9.1 Explorer 10 was the first spacecraft carrying instruments into space to measure the solar wind (Image credit: NASA)

different velocity intervals up to 660 kilometers per second. Weighing in at 35 kilograms, Explorer 10 was launched on 25th May 1961, and surveyed the sky on the night side of the Earth. Although it was not clear at the time, the spacecraft never completely escaped from the Earth's magnetic field proper. Despite this drawback, the plasma flux of 4×10^8 ions per square cm per second recorded by Explorer 10 was similar to that measured two years earlier by Luna 2. Having presumed that the plasma mostly consisted of protons the measured ion speed ranged from 120 to 660 kilometers per second, with a mean of about 300 kilometers per second flowing directly away from the Sun. These data bolstered Biermann's theory. That said, the measured plasma density varied from about 6 to 20 protons per cubic centimeter, which was significantly lower than that predicted by Biermann. Unfortunately, Explorer 10's measurements displayed a considerable amount of variability, owing to the nature of its orbit, which placed it in a region between the magnetopause and the bow shock. As a consequence, some of the scientific community still had doubts about the validity of Biermann's and Parker's theories. To break the deadlock, either more data had to be gathered from space on the dayside of the Earth and/or from interplanetary probes.

While the team at MIT was continuing with their work, Marcia Neugenbauer and Conway Snyder of JPL produced a prototype plasma spectrometer in 1959 to measure the energy spectrum of both electrons and protons derived from the solar wind. Confident that it would work, they successfully lobbied NASA officials for its inclusion in an early lunar or interplanetary probe. Their first available launch came with the unmanned US probes Rangers 1 and 2, which were both planned for a lunar flyby in 1961. Unfortunately both craft were unable to leave their Earth parking orbits because of a technical glitch with their launch vehicles. The next opportunity to test the plasma spectrometer came with the Mariner 1 and 2 space probes, which were launched towards Venus in 1962. The launch of Mariner 1 proved a failure, but Marnier 2 achieved a successful launch on 27th August 1962.

Neugenbauer and Synder's plasma spectrometer revealed that the velocity of the solar wind, which always blew outwards from the Sun, was quite variable in intensity. In general the recorded velocities fell in the range from 400 to 700 kilometers per second, but occasionally it reached very high velocities up to the 1250 kilometers per second- the maximum that could be measured with the then available technology. The positive ions in the wind were generally assumed to be protons, at a measured density of 1 to 5 protons per cubic centimeter, but some helium nuclei (alpha particles) were also detected. What's more the velocity peaks seem to correlate with peaks in magnetic activity and which recurred with a period of 27 days. Thus, for the first time, these solar plasma ejections could be clearly and unambiguously linked to magnetic storms on Earth. Not content to rest on their laurels, Neugenbauer and Synder began a study to identify the M regions on the Sun responsible for generating these storms, but the lack of enough quality data precluded any such analysis to be conducted.

While the solar wind results derived from Mariner 2 were still being analyzed, NASA pushed ahead with the launch of a completely novel type of spacecraft. Called the Interplanetary Monitoring Platform, IMP-1, the 62 kilogram probe was launched on 27th November 1963. One of the principal investigators assigned to this mission was Norman Ness of NASA's Goddard Space Flight Center, who was responsible for the satellite magnetometer. The other principal investigator assigned to IMP-1, John Wilcox of the University of California at Berkeley, liaised with the staff of Mount Wilson's solar observatory. Wilcox compared and contrasted Mount Wilson's measurements of solar magnetic fields with the data garnered by the IMP-1 spacecraft.

IMP-1 was the first of a series of ten IMP spacecraft planned to investigate the interplanetary environment over the course of a complete (11-year) solar cycle. Of these the first three had identical specifications. The experiment payload of IMP-1 consisted of three magnetometers, four cosmic ray experiments and four solar wind particle experiments. These spacecraft were equipped with a better rubidium vapor-type magneto meter that flew on board the Explorer 10 probe. The instrument was mounted at the end of a 1.8 meter axial boom to minimize potential interference from the spacecraft proper. Another 2-meter long boom carried two fluxgate magnetometers to measure the direction of the solar wind. The rubidium magnetometer, which had a measurement range of from 0.1 to 1000 gammas, was designed to measure the intensity of the interplanetary magnetic field.

Mission controllers wished to place the IMP-1 spacecraft into a highly eccentric, six day orbit with an apogee of 278,000 kilometers on the sunward side of the Earth, but after a launcher malfunction, its orbit had to be re-adjusted into a four-day orbit with an apogee of 198,000 kilometers, but at least it was still on the sunward side of the Earth. At first the spacecraft spent about three quarters of its time during each orbit in the region of interplanetary space beyond the Earth's bow shock, but as time progressed IMP-1 spent progressively less time during each orbit in interplanetary space, owing to the changing orientation of the Sun-Earth line as the earth orbited the Sun. By mid-February 1964 the time in interplanetary space had reduced to zero, but by then the Sun had undergone almost three full revolutions and providing a wealth of new solar data in the process.

As discussed above, Parker had shown as early as 1958 that the Sun's magnetic field lines spiral out from the Sun in a pattern similar to that produced by a rotating lawn sprinkler. He further calculated that the magnetic field lines at the Earth's distance from the Sun should make an angle of about 45 degrees to the Sun-Earth line. Using fresh IMP-1 magnetometer data, Norman Ness was able to confirm that the magnetic field lines did indeed make such an angle to the Sun-Earth line. Curiously, Ness was surprised to discover that every few days the magnetic lines suddenly changed direction by as much as 180 degrees. This prompted Ness and Wilcox to initiate a detailed study of the correlation between the spacecraft measurements and those obtained from ground-based telescopes at Mount Wilson. The study revealed that the interplanetary field's polarity was strongly correlated with that of the Sun's photosphere at the Sun's equator with a lag of about 4.5 days. Furthermore, this implied an average solar wind velocity of about 380 kilometers per second which was consistent with that measured by IMP-1's plasma trap. What was now clear was that the interplanetary magnetic field was an extension of that of the Sun, changing as that on the Sun changed.

After all the data was analyzed covering after nearly three solar rotations Ness and Wilcox found that the abrupt, 180 degree changes in field direction seemed to occur at about the same place relative to the Sun's surface during each revolution. The only exception to this trend was observed on 2nd December 1963, when a solar storm induced the field to change direction suddenly. This led them to suggest that the Sun-generated interplanetary magnetic field is composed of organized sectors of alternately inward an outward pointing magnetic fields. They also found that the passage of IMP-1 across the sector boundaries was not only evident in the interplanetary magnetic field but also I the plasma velocity measured by the spacecraft, and in the geomagnetic index measured on Earth,

Its primary mission now completed, between February and May 1964, IMP-1 was put to use investigating the Earth's magnetosphere bounded by the magnetopause. As a result of such a study, it was discovered that the magnetic tail on the night side of the Earth extended much further than anyone expected- well beyond the satellite's apogee of 32 Earth radii. Field strengths of about 20 gammas were measured inside the magnetosphere, compared with the interplanetary field at the Earth's orbit of about 5 gammas. IMP-1 also uncovered a narrow region in the magnetosphere at local midnight where the field intensity was practically zero. The so-called 'neutral sheet' divided the region to its north, where the Earth's magnetic field lines were pointing sunward, from that to its south, where the field lines were pointing in the opposite direction.

The effect of the Moon on the solar wind was first investigated by IMP-6 (or Explorer 35) which was put into lunar orbit on July 22nd 1967. Unlike many of the planets, the Moon has practically no magnetic field and, as a result, has no bow shock. Instead IMP-6 found that the sunlit surface of the Moon appears to be directly bombarded by the solar wind, which produces a shadow void in the solar wind on side of the Moon not facing the Sun. IMP-6 also conducted important work tracing out the interplanetary magnetic field close to the Moon and the effect of the Earth's magnetospheric tail as it swept past the Moon every 29.5 days.

Cosmic rays are deflected as they interact with the Earth's magnetic field to create so-called secondary cosmic rays. As mentioned previously, as early as the 1940s Forbush had found a slight decrease in secondary cosmic rays at the surface of the Earth during solar maximum. The average change was rather small though, typically only 4%, in fact. Over a five year period, between 1967 and 1972, the IMP-6 probe mapped the intensity of primary cosmic rays in interplanetary space, finding that the effect was much more marked, with a maximum typically 80 % above the minimum. The underlying reason for this is that we only detect secondaries from the higher energy cosmic rays at the Earth's surface because the cosmic rays with the lowest energy are completely absorbed by our atmosphere, while the higher energy cosmic rays are largely unaffected by the solar magnetic field. There are cosmic rays of all energies in interplanetary space however, including the lower energy species which are most affected by changes in the solar magnetic field. The Pioneers 10 and 11 probes launched back in 1972 and 1973 also recorded a gradual increase in the average cosmic rays intensity of about 1.5 to 2 % per astronomical unit (AU) as they prepared to leave the solar system.

During the 1960s and early 1970s, two major unsolved problems preoccupied solar physicists. Precisely where on the Sun's surface are Bartel's M regions, the expected source of the high-speed particle streams. Another thing that needed answering was the question of the precise three-dimensional shaped interplanetary magnetic field. Were they really shaped like the segments of an orange, as surmised by Wilcox and Ness, each running in a North-South plane back to the Sun, or were they the intersections of a warped neutral sheet and the ecliptic? Unfortunately, no man-made craft had ventured very far from the ecliptic, so the three-dimensional structure of the interplanetary field could only be inferred from various, indirect evidence.

As early as 1957, so-called coronal holes had been observed at visible wavelengths—which were believed to be gaps in the corona on the solar limb. Towards the end of the 1960s, these observations were extended into the extreme ultraviolet by the Orbiting Solar Observatory (OSO)-4 spacecraft. These observations revealed coronal holes on the solar disk. That said, coronal holes were best seen at X-ray wavelengths, where they show up as dark regions of the Sun running from pole to pole. In 1973 Allen Krieger and Adrienne Timothy of American Science and Engineering Inc., together with Edmund Roel of the University of New Hampshire who showed that there was a correlation between low X-ray intensities at the solar equator and high solar wind velocities measured by the Vela and Pioneer 6 spacecraft. This was the powerful new evidence that coronal holes were indeed Bartel's M regions, which

Fig. 9.2 NASA's Skylab orbiting the Earth (Image credit: NASA)

generate a faster solar wind. Subsequent observations conducted by solar physicists confirmed these findings of coronal holes using observations made on board Skylab, OSO-7 and the solar telescope at Kitt Peak, as well as wind speed data with IMP-4 and 7 and the German spacecraft Helios 1 (Fig. 9.2).

Coronal holes become more prevalent in the period leading up to solar sunspot minimum. As luck would have it, Skylab, equipped with its two high-resolution X-ray telescopes, was up and running just before the 1976 solar minimum. A total of six low-altitude coronal holes, each lasting typically for a number of months, were observed in the Skylab. Furthermore, these holes were shown to be associated with large unipolar magnetic field regions deep inside the Sun's photosphere. In addition, the boundaries of these coronal holes were the sites of magnetic inversion lines, where the field changed polarity. So, not only were coronal holes Bartel's M regions, but Babcock's original idea was also proven to be correct, that is, high energy cosmic rays emanate from unipolar regions where the magnetic field lines run freely into space.

The second problem mentioned above, namely the shape of the interplanetary field in the region of the ecliptic, was solved in 1976 when Pioneer 11 found that the segmented magnetic 'slices' completely disappeared when its heliographic latitude reached 15 degrees or above, affirming the warped sheet model, although the three-dimensional location of this sheet and the morphology of the interplanetary field still remained mysteries. More observations carried out during the solar maximum of 1979 showed that the neutral sheet changed and new, secondary neutral sheets appeared during the time the Sun's magnetic field changed polarity.

As we have seen, during the late 1940s and early 1950s, Friedman's group at NRL had undertaken a number of ultraviolet and X-ray observations using sounding rockets. Then in 1956 Freidman attempted to observe X-rays from solar flares. Unfortunately, because of the transient and short lifetime (typically minutes) of these flares, the Aerobee sounding rockets he relied on simply took too long to prepare for launch. Clearly, Friedmann had to look for a more efficient way around the problem. Solid-fuelled sounding rocket technology was more suitable as the smaller rockets could be launched at a moment's notice; enter the Rockoon.

Consisting of a 4-metre long solid fuelled Deacon rocket, the Rockoon was suspended from a Skyhook balloon that floated up through the atmosphere to an altitude of about 20 to 25 km. The logistics of this mission were as follows; the Rockoon would be launched in the morning, and the team could track its drift in the atmosphere by radar. When a flare was seen, it would fire its rocket motor. A large launch range was required to allow for balloon drift and to cover the likely impact sites of the spent rockets. In the end Friedmann's team used a ship to launch the Rockoon and then follow its trajectory by radar for as long as possible before the rocket was fired. Friedman and his team set out to sea on board the USS Colonial in the summer of 1956 with a batch of ten Rockoons, each one launched every 24 hours.

On the first three days, no flares were seen, but on the next day, two flares were observed but there was no balloon aloft. On the following day the rocket was fired as soon as a flare appeared. At altitudes of 70 kilometers or above, strong X-rays were detected, proving that solar flares do indeed emit X-rays. Friedman also uncovered circumstantial evidence that X-rays can be derived from non-solar sources, but all attempts at finding these during subsequent night time launches were unsuccessful.

Solar flares are not only prodigious emitters of X-rays and radio waves, but they also emit charged particles. This much was shown as early as February 23 1956 when a very bright solar flare was observed in white light against the backdrop of the Sun's photosphere. This flare not only disrupted radio communications on the daylight side of the Earth, as they generally do, but within an hour it had also compromised telecommunications on the night side of the Earth. The latter event could not been caused by X-rays, since, like all electromagnetic waves, they travel in straight lines and would thus not have affected the night side of the planet. Rather it was attributed to flare-generated, charged particles moving in highly cured paths within the Earth's magnetic field. Subsequent observations of similar events conducted aloft from balloons revealed that most of these charged particles responsible for night-time disruptions, and their cause squarely laid at the table of low energy protons in the energy range between 30 to 100 MeV.

The first Lyman alpha image (at 121.6 nm) in the ultraviolet of the Solar disk was also produced in 1956 from a sounding rocket. This was followed in 1960 when the first X-ray solar image was generated using a simple pin-hole camera carried aloft by an Aerobee sounding rocket. This X-ray image, taken over the 2–6 nm wavelength range, was rather crude by today's standards and its resolving power was further compromised because the rocket and its X-ray camera popd during the exposure. Although these first ultraviolet an X-ray images were of limited scientific use, they were certainly good enough to provide the first insights into what was potentially viable using a dedicated solar satellite monitoring our star at ultraviolet and X-ray wavelengths.

The first dedicated solar observational spacecraft- SOLRAD-1- was launched in June 1960. The 19 kilogram satellite controlled by NRL, entered orbit together with a larger Transit 2A navy navigation spacecraft. This was the first time in history that two spacecraft entered the vacuum of space with aboard a single launch vehicle. SOLRAD-1 carried two detectors to monitor the solar environments, one monitoring the Lyman alpha region from 105 to 135 nm and the other at shorter, X-ray wavelengths below

0.8 nm. Unlike many later spacecraft, it did not have a tape recorder on board, relying instead on line-of-sight telemetry with a series of stations on the ground. As luck would have it, SOLRAD-1 happened to be in direct sight of Blossom Point receiving station in Maryland on August 6 1960, when a powerful flare was observed from Earth. At the same time, a series of telecommunication problems were experienced by Earth-bound stations and the recorded cosmic ray intensity was also attenuated. Just four minutes after the visible flare was observed, radio emissions from the Sun were picked up on Earth. SOLRAD-1 reported a rapid increase in X-ray intensity which lasted for a much longer than the radio burst. Indeed, throughout the short period of time that the flare lasted, every conceivable flare-related phenomenon then known was observed.

In parallel with the US-navy run SOLRAD spacecraft, Ball Brothers were commissioned to design and build a much larger (200 kilogram) Orbiting Solar Observatory (OSO-1) spacecraft by NASA. Launched by a Delta rocket on March 7 1962, OSO-1 consisted of a nine-sided equipment platform that was spun up immediately after entering the space environment in order to provide gyroscopic stability. OSO-1 also carried a de-spun, fan-shaped platform partially covered in solar cells (Fig. 9.3).

Fɪɢ. 9.3 Artist's impression of OSO-1 in Earth orbit (Image credit: www. space.skyrocket.de)

A total of eight experiments were mounted on the spinning equipment platform and another five on the de-spun platform. The spacecraft measured short-wavelength visible radiation between 380 nm and 480 nm, ultraviolet radiation between 110 and 125 nm (including Lyman alpha), X-rays in three wavelength bands between 40 and 0.1 nm an gamma rays of various energies between 0.05 and 500 MeV. OSO-1 instruments were also able to measure neutrons, protons, electrons and even micrometeoroids. In all, a total of six OSO spacecraft were successfully launched in this first ambitious space program ending with the launch of the 290 kilogram OSO-6 in 1969.

Though OSO-1 was anticipated to last about six months, its mission lifetime only lasted 76 days before a spin control system failure resulted in only intermittent operation. That said, during its short lifespan OSO-1 observed more than 75 flares and sub-flares and mapped the gamma ray sky, as well as monitoring the inner Van Allen belt at an altitude of just 550 km. Five years later, one of its successor spacecraft, OSO-3, managed to capture the high-energy spectrum of a flare that meant that it emanated from super-hot plasma with an estimated temperature of 30 million K!

As we have seen, our Sun had been known for some years to emit intense but relatively short-lived bursts and across many frequencies, especially during solar maximum. Solar astronomers divided these radio pulses into four categories I through IV. Type I radio bursts were strongly correlated with sunspots, rather than flares. Type III bursts were the first to arrive at the Earth after a flare was first observed, which were of medium to low frequency(<500 MHz) but which also reduced in frequency rapidly as a function of time (about 20 MHz/s). Type II, on the other hand, arrived later and displayed a lower frequency (<100 MHz), and which gradually faded with time at a rate of about 0.1 MHz/s. Finally, type IV bursts, which are quite often associated with their type II counterparts, emitted over a broad, high frequency range, and were likely derived from charged particles that are accelerated in a magnetic field, that is, by synchrotron radiation.

While ground-based radio telescopes helped to characterize these radio bursts in the 1950s and 60s, spacecraft have had to be used to cover the very important frequencies, as the ionosphere is not transparent below about 10 MHz. By the mid-1960s Canadian scientists had detected type III radio bursts at frequencies between 0.7 and 10 MHz using the Canadian designed Alouette 1 and 2 spacecraft. Similar observations were conducted within a couple years by scientists using OSO-3, as well as the Russian Zond 3 spacecraft. It was known that the radio frequency of these bursts seemed to correlate with the electron density at the place of their origin. By determining how the ambient electron density varies with distance from the Sun, astronomers were able to estimate the source velocity relative to the Sun, by measuring the rate of change of its radio frequency with time. By this means, the velocity of particles producing type III radio bursts was determined to be about 30 to 40 % of the speed of light. These velocities were subsequently confirmed in August 1968, when NASA scientists were able to directly determine the velocity of these type III associated particles by tracking the source using the recently launched RAE-1 spacecraft. This new data found that their velocity seemed to be constant at about 35 % of the speed of light extending as far out as 40 solar radii from the Sun. Follow-up observations undertaken with the IMP-6 spacecraft determined the source of one type III burst which allowed it to be traced as far back as the Earth's orbit, at which time its velocity had decreased by half. As this source was being analyzed, IMP-6 also observed a burst of solar electrons travelling at a significant fraction of the speed of light.

Bizarrely, some electrons emitted by type III bursts occasionally return to the Sun in an event called a 'U' burst. Such an event occurred on November 29th 1956, when a burst was observed at Fort Davis, Texas, and over a period of 8 seconds its radio frequency was seen to attenuate from about 175 MHz to 125 MHz after which time it was observed to increase again! A similar but larger event was observed in the mid-1960s by the RAE-1 spacecraft when the radio frequency from another type III burst was seen to decrease from about 5 to 1 MHz and then rise again to 3 MHz over a period of only seven minutes. This was interpreted as being caused by a stream of relativistic electrons that spiral outwards to about 35 solar radii from the Sun before falling inwards again, along the magnetic field lines, back to the solar disk.

In August 1972, a series of intense solar flares were recorded, and all derived from the same active region of the solar disk. A trio of flares were seen on August 2, and two others were recorded on the 4th and 7th. Of these, it was the August 4th flare that proved most significant OSO-7 discovered that gamma rays were being emitted and so, by inference, must have been derived from a very energetic collision, involving the mutual annihilation of electrons and positrons producing a peak at 0.5 MeV. Particle physicists have long known that mutual annihilations like these must take place on very short timescales, the positron must have been produced in the solar flare, possibly by the decay of short-lived isotopes of carbon, nitrogen and oxygen, which had also been produced in the flare. Yet another strong gamma ray emission line was at 2.2 MeV, is produced following the production of deuterium by the collision of a neutron and proton in the flare.

In these and other ways, dedicated solar observatories transformed solar astronomy from a fringe discipline to the center of attention in the world of astrophysics. Flares were not only able to accelerate electrons, as seen in type III radio bursts, but protons also, which manifest a great many forms of electromagnetic radiation and participate in various nuclear reactions. Calculations showed that the observed effects of solar flares, particularly the shape of the X-ray continuum spectrum, could only be explained if the temperatures exceeded 100 million K, with the energy deriving from the mutual annihilation of particles with their antiparticles much in the same way as the collisions that occur in particle accelerators.

On the afternoon of August 7th 1972, the most intense solar flare of the series was observed which generated X-rays so intense that the instruments on SOLRAD-9 and 10 were completely saturated. Gamma rays were again detected, and telecommunications compromised on Earth. High energy solar cosmic rays were observed shortly afterwards and Pioneer 9 detected lower energy particles. A type II radio burst was also detected at Fort Davis by a radio. The IMP-6 spacecraft recorded the reduction in radio frequency and over the next 24 hours tracked it all the way out to the Earth's orbit, by which time the radio frequency had reduced from about 180 MHz to just 30 MHz with a steady velocity of 1270 km/s. When the source arrived at the Earth shortly before midnight (UT) on August 8, it produced a geomagnetic storm, thereby confirming the hypothesis that terrestrial magnetic storms, which include energetic aurora displays, result from a shock wave that originates from a solar flare, while the type II radio burst representing the front of the shock wave itself.

It is the Sun's chromosphere that solar flares generate the electrons and shock waves described above, but they then have to travel through the solar corona to reach the Earth. The size and shape of the corona had long been known to vary according to the phase of the solar cycle, being more symmetrical around solar maximum and less so near solar minimum. That said, it was not until well into the 1960s that it became possible to attempt to correlate X-ray coronal activity with the large coronal streamers, which are seen in white light to stretch for several solar radii above the chromosphere. From that time onwards, it was generally understood that these enormous white-light streamers are anchored in X-ray bright patches seen on the solar surface.

While OSO-7 was monitoring the white light solar corona, an exceptionally bright coronal streamer was picked up by a solar telescope observing in the light of hydrogen alpha at Carnarvon, Australia. The same telescope recorded a faint drizzle of hydrogen gas in the space immediately around the streamer. A radio burst was detected at Culgoora, Australia, at the same time and other radio observatories based in the Philippines and the USSR recorded similar bursts. When OSO-7 turned its instruments on the Sun, its instruments showed several clouds of HII (ionized atomic hydrogen or protons) leaving the Sun by means of powerful coronal mass ejections. Being 300,000 km or so in diameter, and having temperatures of the order of about 1 million K or so, these clouds were found to be leaving the Sun at upwards of 1000 kilometers per second. The simplest explanation for this was that the ionized gas cloud was being ejected from a solar flare just beyond the visible edge of the solar limb. 84 hours later the Earth's magnetosphere was awash with low energy charged particles, and shortly thereafter Pioneer 6, then in solar orbit, recorded an increase in high-energy protons. Thankfully, the most energetic HII clouds had not intercepted the Earth.

Prior to the initiation of radical new post-Apollo missions, in early September 1965, NASA administrators had established the Apollo Applications Program in order to employ Apollo-designed hardware for ambitious manned Earth-orbiting and lunar surface missions. This program set up under the Apollo Extension System, was to investigate how to use hardware developed for the Apollo program to construct a manned Earth-orbiting laboratory. By 1966 plans for three orbital workshops were forwarded which included plans to construct them in orbit. 26 Saturn IB and 19 Saturn 5 launches were planned in this manned program with first launch scheduled for April 1968. These projects were to go ahead in parallel with the new manned lunar program following the approved Apollo lunar missions.

As with so many other previous missions, budget cuts saw a progressive reduction in the Apollo Applications Program in the years that followed, with the lunar elements being cancelled in 1968. In addition, the six Earth-orbiting workshops originally planned were mothballed and the number reduced to just one by 1969. This remaining orbital workshop, which was called Skylab, was launched on a Saturn 5 rocket on May 14 1973. Between May 25 1973 and February 8 1974 the orbiting space station was occupied by three crews of three men occupying it for a total of 171 days.

Skylab stands as a lasting testament to the way in which hardware designed for one space mission could be adopted by another. The orbiting laboratory consisted of a modified S-IVB from the third stage of a Saturn 5, together with a multiple docking adapter module at one end to allow access to the Apollo Telescope Mount (ATM) and the Apollo Command and Service Module (CSM). The Saturn V carried aloft the modified S-IVB, the docking adapter and the ATM were all, while the CSM was launched by a Saturn IB rocket carrying a three man crew just a few days later. Extensive modifications had to be undertaken on the original S-IVB stage to allow the crew to live comfortably on board Skylab, as well as creating the laboratories for the astronauts to work in and equipping it with an airlock module. This came to be known as the orbital workshop (OWS). The power to allow Skylab to operate Skylab was derived from two large solar array panels affixed one on either side of the S-IVB stage, and a further four solar arrays attached in a windmill-like supplied electrical power to the ATM. The spacecraft's overall length measured in at 36 meters, with a total mass of about 90 tons.

Most of the solar instruments on board Skylab were situated on the ATM. One of the big advantages of this manned astronomical observatory, compared with the previous un-manned missions was that most of the images could now be stored on film and analyzed in situ instead of being transmitted to the stations on the ground. Film also allowed the transmission of far more information back to Earth than could have been possible by previous technologies. What is more, Skylab's great size and mass enabled much more power and space to be made available to the experimenters than in any Earth-orbiting solar observatory spacecraft previously designed.

After an almost faultless countdown on May 14 1973, after just 63 seconds it encountered a slight vibration as the Saturn 5 vehicle passed through the point of maximum aerodynamic pressure. Shortly afterwards, the launcher's protective nose cone was then ejected and the rocket motors shut down, The AT was then popd through 90 degrees to make it perpendicular to the S-IVB stage. The ATM's four cross-shaped solar arrays were successfully deployed, but the two solar cell panels attached to opposite sides of the S-IVB stage could not be deployed successfully. The glitch was traced to a heat shield and one of the solar panel wings, which had fallen off during the launch causing the heat shield to inflict damage to the other solar panel in the process. As a consequence, the power supply to Skylab was severely compromised and this resulted in rapid heating of the interior of the spacecraft to 88 C. Subsequent attitude maneuvers brought the temperature down to 43C. These temperatures were still unacceptably high for human habitation the launching of the astronauts had to be postponed while NASA officials frantically designed a new deployable sunshield to fix the damaged Skylab. Finally, on May 25 1973 the astronauts Charles Conrad, Paul Weitz and Joseph Kerwin were successfully carried into Earth orbit and managed to fix the spacecraft and operate it for the next four weeks.

With the heat shield successfully repaired and operating, Skylab's internal temperature fell off to comfortable levels, and by May 30 the astronauts were able to initiate routine work. While all of the ATM's experiments worked well, some of the workshop experiments were aborted because part of the sunward side of the spacecraft had to be covered by the makeshift sunshield. By June 7 the astronauts succeeded in successfully deploying the damaged wing, which provided an additional 3000 W of power, but this was still depressingly short of the 12.4 kilowatts that had been expected from both wings. And although the ATM solar arrays provided the expected 10.5 kilowatts of power, Skylab still suffered from a severe power shortage, placing severe restrictions on the number of experiments that could be successfully completed.

Despite these significant setbacks, the astronauts on board Skylab were able to use its state-of-the-art scientific instruments to improve and refine our understanding of the Sun over a period of months. These included pioneering observations on coronal holes, as well as regular monitoring of active regions within the chromosphere which revealed a strong linkage between the two phenomena. These arc-like features, called coronal loops, which account for nearly all the electromagnetic energy emitted by the corona, were also found to link areas of opposite polarity in the photosphere.

Visible images of the Sun's outer atmosphere during an eclipse going all the way back to the nineteenth century showed prominent plumes emerging and moving outwards from its polar regions. Furthermore, these plumes were most energetic around solar maximum, but their origin was not clear. Skylab provided the answer to this question when it captured ultraviolet images of the polar regions which showed that there were small, bright patches associated with strong magnetic fields seen in the otherwise dark coronal holes These bright patches being the source of the plumes (Fig. 9.4).

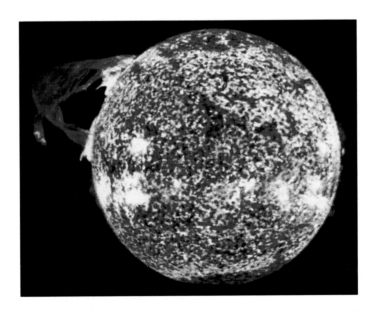

FIG. 9.4 A solar prominence imaged by instruments on board Skylab, captured on December 19, 1973 (Image credit: NASA)

These bright X-ray spots, which were about ten times smaller than the aforementioned arch-like features, were actually seen by G.S Vaina, who employed a sounding rocket in 1969, but their detailed study could only be undertaken by Golub, Vaina and others using the greatly improved instruments on board Skylab. After the astronauts analyzed images taken by one of the X-ray telescopes, and typically found about 100 bright patches on each image distributed fairly evenly across the solar disk. This random distribution was distinctly different from sunspots however, which were generally confined mid- and lower solar latitudes. This came as a great surprise to astronomers Moreover, the longevity of the bright points were of the order of about 8 hours—much shorter than sunspots—and a few of these spots displayed exceptional brightening over periods of just a few minutes before fading away just as quickly. These bright spots were also found in bipolar regions, and seemed to be associated with localized magnetic fields of opposite polarity In addition, the total magnetic flux contained in these bright spots were found to be similar to that contained in sunspots and active regions. Later work carried out by ground-based radio telescopes, such as the Clark Lake Radio Observatory, showed them to be associated with type III radio bursts which in turn were known to be connected with solar flares.

Skylab also carried out a detailed examination of over 100 coronal transients, which are typically characterized by major, rapid, events occurring in the corona. Of these, 77 were thought to be coronal mass ejections (CMEs) like those discovered by OSO-7. Most of these CMEs, which sling billions of tons of material outward from the solar atmosphere with velocities up to 1200 km/s, were strongly linked with eruptive prominences, while others were associated with flares That said, it was not entirely clear whether these CMEs caused the prominences and flares to appear, or vice versa. CMEs are however, much larger than prominences, often becoming larger than the Sun itself! And while the matter within them is very tenuous, a large, fast-moving CME can be as energetic as a large flare (with energy of about 10^{25} J).

The next major solar observation spacecraft to be launched was Solar Max, a 2400 kilogram satellite which blasted off in February 1980. Solar Max's mission was to primarily study solar flares in particular, around the time of solar maximum. The spacecraft carried seven experiments including the first telescope

Fig. 9.5 Solar Max: artist's impression of the spacecraft in orbit (Image credit: NASA)

designed to image the Sun in high energy X-rays (the so-called Hard X-ray Imaging Spectrometer) (Fig. 9.5).

Its other instruments included an ultraviolet spectrometer a polarimeter (UVSP), a white light coronagraph/polarimeter (C/P), a gamma ray spectrometer (GRS), and a total solar irradiance monitor. A few of these instruments had no imaging capability whatsoever, but were designed to measure the total energy emitted by the Sun over different wavebands, whereas others were designed specifically to image active areas of the disk as well as eruptive flares. Finally, the C/P was dedicated to obtaining images of the solar corona at visible wavelengths.

Unlike any solar observatory before it, the Solar Max probe was designed to be retrieved at the end of its useful lifetime in December 1982 by the Space Shuttle and returned to Earth for refurbishment and repair. But in November 1980 all but one of the spacecraft's gyros malfunctioned, with the resulting loss of its fine pointing capability. As a result, Solar Max could no longer image the Sun but could still use its other instruments. But all that changed in April 1984 when space shuttle astronauts managed to repair and refurbish Solar Max from Earth orbit. These refurbishments enabled the spacecraft to operate until December 1989 after which it burned up in a blaze of glory in the atmosphere.

   The repair and refurbishment of Solar Max was portrayed by NASA as a vindication of their policy of having a manned Space Shuttle that can undertake in-orbit repairs and maintenance of spacecraft, which was the reason for NASA's publicity of the event. The costs of a Space Shuttle mission (at about $300 million a time) are such, however, that it may have been less expensive to launch a replacement spacecraft with another Delta, although the Shuttle did launch another spacecraft during the Solar Max repair and refurbishment mission.

   Before the launch of Solar Max there were two main theories of the solar flare phenomenon. In one, flares are caused by the re-connection of magnetic field lines in the coronal part of an active region. The energy released accelerates electrons that then spiral down the field lines to the solar transition region and chromosphere. There, the high-energy electrons are suddenly decelerated, giving off short bursts of hard X-rays, radio and gamma radiation in the so-called impulsive stage of the flare process. Some energy released by the electrons also heats up the chromospheric gas, causing it to rise into the almost empty magnetic loop, where it produces soft X-rays. In the alternative theory the flare process starts with the compression and heating of gas in the chromosphere by magnetic fields, with the gas then expanding along a magnetic loop into the corona.

   The earliest images returned by the Hard X-ray Imaging Spectrometer (HXIS) on Solar Max showed conclusively that, hard X-rays are emitted by flares generated at the foot of the magnetic loop. On either side of these magnetic loops, the flares were observed to brighten for about five seconds, which approached the maximum time resolution of the HXIS instrument. Further observations carried out by HXIS and the Hard X-ray Burst Spectrometer (HXRBS) demonstrated that the emission of hard X-rays reached their peak before that of soft X-rays. What is more, the peak in hard X-rays was shown to be considerably sharper than the peak in soft X-rays and extended well into the corona. Curiously, the UVSP and HXRBS showed a remarkable correlation between the intensity variations with time of the ionized oxygen line (OV) at 131.7.nm in the ultraviolet, the nearby ultraviolet continuum at 138.8 nm, and hard X-rays. Ultraviolet emission in the chromosphere and coronal emission lines are not impulsive like that in the transition region lines, however, having a slower rise to maximum, like that of the short X-rays.

In addition to hard and soft X-rays, the OSO-7 spacecraft had shown that some very energetic solar flares emitted gamma rays at energies of 0.5, 2.2, 4.4 and 6.1 MeV caused by electron/positron annihilation (0.5 MeV), the formation of deuterium (2.2 MeV) and e-excitation following the collision of protons with nuclei of carbon (4.4 MeV) and oxygen (6.1 MeV). Prior to the launch of Solar Max, most solar physicists were of the opinion that electrons were accelerated in a solar flare ahead of the more massive protons, the latter being accelerated in a separate, second stage of the flare. But that being the case, hard X-rays ought to have been produced before the proton-associated gamma ray lines, since hard X-rays are produced by electrons. Unfortunately OSO-7 did not have a good enough time resolution in its gamma ray detector to verify this, but the gamma ray spectrometer (GRS) instrument on Solar Max did manage to show that electrons and protons are accelerated simultaneously. In addition, the gamma ray spectrum measured by Solar Max showed a series of peaks at 0.5, 1.3, 1.7, 2.2, 4.4 and 6.1 MeV superimposed on a continuous background caused by the deceleration of electrons, generating so-called Bremsstrahlung radiation in the solar atmosphere, with the novel 1.3 an 1.7 MeV lines being generated by the de-excitation of magnesium an neon nuclei following their collisions with protons.

When an unusually large solar flare was observed in 1980 and again in June 1981 the GRS instrument on board Solar Max detected neutrons about two minutes after it detected the hard X-rays. If the neutrons were emitted at the same time as the hard X-rays, their velocity must have been of the order of 80 % of the speed of light. The best explanation for these high-velocity neutrons was that they must have originated from the collision of protons that had been accelerated to an extremely high energy by the flare.

The data from Solar Max had clearly demonstrated that solar flares initiate with the reconnection of the magnetic field lines in the corona which then accelerate electrons and protons. The electrons spiral along field lines to the solar transition region and chromosphere while protons originate in the photosphere. That said, some very high energy electrons and protons can freely escape from the Sun altogether, and have enough kinetic energy to reach the Earth 150 million kilometer s away. The electrons that spiral down the field lines to the Sun are rapidly decelerated in the transition region and chromosphere, where the emit short bursts of hard X-rays, ultraviolet, type III radio waves and gamma radiation in the impulsive phase of the magnetic loop, raising the temperature of the chromosphere. Its constituent gas then undergoes expansion, when it rises and fills the magnetic loop, emitting soft X-rays in the process. But it also retains gas in the chromosphere at the base of the loop where it strongly emits at ultraviolet and optical wavelengths. In the meantime, the high energy protons have spiraled down the magnetic field lines to reach the photosphere, where they collide with photospheric protons to produce neutrons. These, in turn, are captured by protons to form deuterium, emitting 2.2 MeV gamma rays in the process. Similarly, flare-generated protons impact magnesium, neon, carbon and oxygen nuclei, which then emit gamma ray lines as they settle to a lower energy state. Short-lived isotopes of carbon, nitrogen and oxygen, also produced by flare-generated protons, decay, producing positrons which then interact with electrons in a mutual annihilation process to produce 0.5 MeV gamma rays. Some of the highest energy neutrons produced in the flare are capable of reaching the Earth.

Although the underlying physics of flare production sounds plausible, some display anomalous behavior. In particular, some processes appeared to occur in some flares but not in others. This raised doubts among solar astronomers on the question as to exactly how and where the reconnection of magnetic field lines occurs. It was anticipated that the instruments of higher spatial and time resolution would settle the issue once and for all.

Astronomers also used Solar Max's white light coronagraph to observe the corona during a period of high solar activity between 1980 and 1989, and during a period of low activity which occurred around 1986. At its peak, one or two CMEs were observed per day, and like coronal streamers, they showed up at most solar latitudes around solar maximum. At solar minimum, on the other hand, CMEs were found only at low latitudes, just like coronal streamers.

One of the great accomplishments of Solar Max was that it enabled astronomers to study the relationship between CMEs and eruptive prominences. CMEs were often seen to consist of a bright loop of material moving away from the Sun with a dark cavity behind it, followed by a second loop that sometimes turned out to be an eruptive prominence. For a long time, it was believed that eruptive prominences probably triggered CMEs, but the Solar Max data suggested that this might not be case, as many CMEs were recorded without associated prominences. However, the CMEs that were associated with flares were also found to occur a few minutes before the flare itself, suggesting the CMEs were the likely trigger.

Since the dawn of the nineteenth century attempts had been made to correlate variations in sunspot numbers with fluctuations in the Earth's climate, on the basis that the thermal energy output from the Sun would vary as the sunspot number varied. Astronomers were divided, however, as to whether the Sun's heat output, measured by the so-called solar constant, would increase or decrease as the number of sunspots visible on the Sun varied.

This arena of solar research received a boost in the middle of the nineteenth century after the discovery of the 11 year sunspot cycle. The Swiss astronomer and mathematician, Rudolf Wolf, carried out original historical research in this area by collating sunspot records back to 1700 and comparing them with meteorological records in Europe over the same period. Wolf's research failed to uncover any hard and fast correlations. The best correlation that could be found was a possible link between the Little Ice Age in Europe in the seventeenth century and the so-called Maunder Minimum between 1645 and 1715, when there were virtually no sunspots recorded on the solar disc.

At the beginning of the twentieth century the American astronomer and founding father of the science of dendrochronology, A.E. Douglas, investigated climatic data for Arizona going back 1900 years by measuring the widths of tree rings. He thought that he had detected an eleven year cycle in the widths of the rings, but over the period from 1650 to 1740 this pattern could not be found. In 1922 Maunder pointed out to Douglas that this period was similar to that of the Maunder Minimum, when sunspots were virtually non-existent on the Sun, thus showing a linkage between sunspots and climate. This possible correlation caused an upsurge in research activity attempting to link climate with sunspots but to no effect. In fact, it was found that tree ring data around the world was inconsistent, scuppering any chances of linking global climate with sunspots.

Sunspot records were notoriously unreliable before about 1650, so to examine possible longer term correlations between climate and solar activity another way of estimating solar activity was required that could provide data for earlier centuries. As we saw previously, Forbush had found a decrease in cosmic rays at the surface of the Earth around solar maximum back in the 1940s. What's more, cosmic rays were known to transform atmospheric nitrogen into radioactive carbon-14, which has a half-life of 5730 years. So a decrease in cosmic rays around solar maximum should be mirrored by a decrease in carbon-14 on the Earth during the same period. Since the more common isotope of carbon, carbon-12, which is assimilated by trees in the form of carbon dioxide at the same rate as carbon-14, the ratio of carbon-14 to carbon-12 in tree rings, might give an indication of solar activity back over time.

But it wasn't that simple. For one thing, the effect would be delayed by about twenty years, however, as the radioactive carbon-14 takes this long to diffuse from the upper atmosphere where it originates to the troposphere. When this radioactive carbon was measured in tree rings, the Maunder Minimum was clearly seen as a period of high radioactive levels, as expected, showing that radioactive carbon levels in tree rings did indeed provide a good indication of solar activity. De Vries also found in 1958 that there was another peak in the radioactive carbon from about 1460 to 1540, now called the Sporer Minimum, which was eventually found to correlate approximately with a series of unusually cold winters in Europe.

Attempts to directly measure variations in solar heat output were hampered by variations in the Earth's atmosphere, but with the advent of sounding rockets and satellites, a way forward was found. After all these years, there was a real hope of measuring variations in the solar constant directly which could then be compared with variations in the Earth's climate. Solar researchers were also interested to know just how the solar constant varied with sunspot number. The first technical hurdle to overcome was how to build a precision radiometer that was small enough to be launched on a sounding rocket or spacecraft.

The incentive to develop accurate radio meter s to measure the solar constant in the early years of the American space program was limited more by engineering problems than anything else. NASA had found that spacecraft surface temperatures in orbit didn't match pre-launch predictions based on ground simulation tests. As a result, any differences recorded could have been due to errors in the solar constant used, inaccurate radiometer s in the space simulation chamber, poor thermal modeling, or in-orbit anomalies in the thermal properties of the surface materials making up the spacecraft. JPL scientists began to conduct investigations into this problem, and in 1964 commissioned Eppley Laboratory of Newport, Rhode Island, to measure the solar constant in order to determine once and for all if the correct value had been used in the ground simulation tests.

A twelve-channel radiometer was built by Andrew Drummond's group at Eppley, based on thermocouple detector technology, to re-measure the solar constant. It was flown in 1967 on the X-15 rocket aircraft at an altitude of about 80 km, and measured a solar constant of 1361 W/m^2. This was about 2.5 % lower than that previously used by officials working for the American space program, and it went some way to explaining the observed in-orbit temperature discrepancies.

As well as the research carried out at the Eppley Laboratory, Joseph Plamondon of JPL installed a cavity radiometer on the interplanetary spacecraft, Mariners 6 and 7, in 1969 to monitor their thermal environment while travelling to Mars. And while these radiometer s as those fashioned by Eppley, it was hoped that they were sufficiently stable to measure variations in the solar constant with time. This was not to be however, since the instrumental variations they recorded were too large and concealed any true solar variations.

Prior to the launch of Mariner 6 and 7 Plamondon had correctly surmised that more accurate radiometer s were required to address this question, explaining why Eppley had received the contract to develop such instruments. But in 1965 James Kendall, an employee of Plamondon at JPL, had constructed a much more accurate cavity radiometer, but this was not adapted to work in the vacuum of space. Richard Willson, a scientist working at JPL, modified Kendall's cavity radiometer to work as spacecraft instrument, and flew his prototype on a high altitude balloon in 1968. Kendall's radiometer measured a solar constant of 1370 W/ m^2, but the absolute uncertainty was still too high at ±2 %, although the result was only 0.7 % higher than the value obtained by the Eppley radiometer.

While this important work represented significant progress, it was no until almost a decade later that a new generation of Eppley radiometer s was to be flown on Nimbus 6 in June 1975, which carried an Eppley thermopile radiometer as part of an Earth Radiation Budget Experiment. Intriguingly, this radiometer produced a solar constant of 1392 W/m^2, with a much smaller absolute error of ±0.2 %, which was 1.5 % higher than the then-accepted value of 1370 W/m^2. Of course, there was no discounting the possibility that this result represented a genuine increase with time, but it needed verification. To this end, NASA launched an Aerobee sounding rocket on June 29, 1976 carrying a suite of five radiometer s, all identical Eppley thermopile radiometer s, all of which were calibrated against a Kendall cavity radiometer. The values they garnered for the solar constant threw the scientific community into a spate of confusion as the two thermopile radiometer s, one on the Nimbus 6 spacecraft and one launched on a sounding rocket, gave identical solar constants of 1389 W/m^2, but the four cavity radiometers yielded values between 1364 and 1369 W/m^2! Because the four cavity radiometers, which were based on no less than three different designs, had produced values that agreed with each other to within ±0.2 %, it now looked as though there was something awry with the thermopile radiometer carried by Nimbus 6, but no one could be absolutely sure.

Following its acceptance in 1970 as the instrument for measuring the International Pyrheliometric Standard, Eppley Laboratory started producing Kendall cavity radiometer s under license. When the time came to design a radiometer for Nimbus 7, Jon Hickey, who had taken over from Andrew Drummond at Eppley, proposed an Eppley-Kendall cavity radiometer as part of the payload for the Nimbus 7 orbiting observatory. The spacecraft was launched in October 1978. Over the next six months, the average solar constant was measured as 1376 W/m^2. Hickey erroneously believed that this represented a real increase from the value measured by the Cavity radiometers on the June 1976 rocket flight.

What's more, the maximum daily deviation from the mean over this half year period was 0.14 %, but in August 1979 a drop of 0.36 % was recorded during October 1979. Although these dips took place at a time of high solar activity, they could not, unfortunately, be unambiguously linked to a definite source on the solar surface.

During the early 1970s, a renewed interest in finding a connection between solar activity and global climate was rekindled by American scientist John Eddy. As we have seen, the eleven year solar cycle had already been exhaustively researched as far back as the early eighteenth century, but many astronomers expressed skepticism that sunspot data earlier than that, which in particular showed the Maunder Minimum from 1645 to 1715, was reliable and in spite of the fact that a string of noted astronomers of the ilk of Johannes Hevelius, John Flamsteed, Jean Picard and Giovanni Domenico Cassini had been keen observers of the Sun throughout the period of the Maunder Minimum and had almost unanimously reported that sunspot numbers were very low or even non-existent, and in spite of the carbon-14 data from tree rings discussed earlier. But the fruits of Eddy's research uncovered more strong evidence for the Maunder Minimum and as a result of his efforts, managed to convince most people that it was a real phenomenon.

In 1976, Eddy produced an analysis of Carbon-14 tree ring data going back to around 3000 BC, showing that the Sun went through a number of periods of high and low sunspot activity, which he was able to correlate with major climate changes, like those of the Maunder and Sporer Minima. This led him to conclude that the carbon-14 data not only showed changes in sunspot activity but also changes in solar heat output, with low sunspot numbers producing a low solar constant and vice versa. Unfortunately, there was still no data to show that the Earth's weather followed the eleven year sunspot cycle. This did not concern Eddy; however, as he thought the sunspot/climate correlation could only be long term.

In the 1970s, whilst Eddy was looking at long-term solar variations, Peter Foukal and Jorge Vernazze were analyzing pyrheliometer data, which measures solar irradiance, collected by Charles Greely Abbot and others at the Smithsonian Astrophysical Laboratory between 1923 and 1952, looking for a possible correlation between the solar constant and the number of sunspots and faculae. Eventually, after a detailed statistical analysis of the 11,000 daily measurements, Foukal and Vernazze concluded in 1979 that sunspots reduce the solar flux while faculae increase it, producing a net short term variability of about 0.1 % in the solar constant. Such small effects were difficult to disentangle from other influences, however, particularly from that of the Earth's atmosphere, whose transmission characteristics could possibly be influenced by solar activity. So their conclusions were generally treated with scepticism by other astronomers. Only measurements above the Earth's atmosphere would be conclusive. Unfortunately the Mariner 6 and 7 radiometer measurements in 1969 were, as mentioned above, not stable enough to show any reliable changes in the solar constant with time. Likewise, the Nimbus 6 results of 1975 and 1976 were also thought to be too unreliable, but in 1979 Nimbus 7 had shown the first short-term correlation of the solar constant with solar activity, although the exact source of the energy reductions was unclear. Nevertheless, John Hickey's Nimbus 7 radiometer had provided a way forward.

While these advances were taking place, Richard Wilson of JPL had his proposed ACRIM radiometer accepted for flight on Solar Max, which, as we saw, was launched in February 1980. That radiometer clearly showed a 0.15 % dip in the solar constant between April 4 and April 9 and a 0.08 % reduction between May 24th and May 28th. Furthermore, both reductions were clearly associated with large sunspot groups that were crossing the central meridian of the solar disk. Further observations revealed that the average solar constant over the first six months of Solar Max was 1368.3 W/m^2 ± 0.1 % with a short term instrument stability of just ± 0.002 %. Wilson's instrument was clearly state-of-the-art for its time.

But if sunspots were causing a small drop in solar irradiance, where did that surplus energy go? The Sun was still producing a constant amount of energy in its core, so that energy blocked by sunspots must either have been stored in the Sun's convective zone below the photosphere and re-emitted later, or it could have been emitted immediately by bright faculae in the vicinity. If it was stored it could either be emitted uniformly or in localized regions by bright faculae. If the emissions of the stored energy were truly global, this could take hundreds of years to appear, whereas if the emission was local it could happen over very short timescales of one or two months.

In 1986, California State University astronomers, Chapman, Herzog and Lawrence analyzed an active region over its entire lifetime and calculated that the facular emission accounted for between 70 % and 120 % of the energy blocked by the sunspots. If the result were 100 % then this would lend credence to the local re-emission theory, but with the large uncertainty this could still support the idea that the faculae were emitting energy stored up years before by other sunspots. If the solar constant closely follows the sunspot cycle, however, the delay in the emission of that surplus energy would, at most, be of the order of months.

After the repair of Solar Max in April 1984, solar physicists expressed considerable interest in comparing the new and accurate solar irradiance data for 1984, just before sunspot minimum, with that of 1980, when a sunspot maximum occurred. Although Solar Max's fine pointing control capability had been after November 1980 to April 1984, its solar irradiance data would still be useable, although admittedly, it was of lower quality the data garnered before the failure. In 1985, using new Solar Max data, Wilson, Hudson, Frohlich and Brusa, showed that there had been a 0.02 % reduction in the solar constant per year in the solar constant in the five year period between 1980 and 1985. A similar reduction had also been observed using the Nimbus 7 radiometer, and so there was little doubt that the data was real. By 1988 however, Wilson and Hudson, reported that the reduction in irradiance had leveled out, and had begun to rise again as the sunspot cycle had passed through its minimum at the end of 1986. Shortly afterwards a similar trend was reported by Hickey and Kerr using both Nimbus 7 and the Earth Radiation Budget Satellite launched by the Space Shuttle Challenger back in October 1984. So although the measured solar constant decreased by the passage of individual sunspot groups across the solar meridian, it began to increase as the Sun approached sunspot maximum. Thus, Eddy's ideas had shown to be correct in associating small numbers of sunspots with a low solar constant, although to be fair he had surmised that this relationship only held over very long periods, and not over the relatively short 11 year solar cycle. But this raised another interesting question In particular, if the individual sunspot groups were the root cause of the observed reduction in the solar constant, was the increase in sunspot number around solar maximum causing a commensurate increase in the solar constant? Or was it due to the faculae which emitted more energy, on average, than that blocked by sunspots?

In important work performed throughout 1988, Foukal and Lean managed to correct the solar irradiance measurements over the period 1981–84 derived from Solar Max and Nimbus 7 for the effect of sunspots. The scientists then compared the resulting curve with that obtained from a specially devised solar faculae index. The match turned out to be excellent, showing that faculae over-compensate for the effects of sunspots.

Oscillations on the solar surface taking place over a period of just five minutes were discovered by Robert Leighton, Robert Noyes an George Simon at the Mount Wilson Observatory back in 1960, and in 1973 Robert Dicke of Princeton University managed to show that this was not a localized effect but instead represented global vibrations of the Sun. Solar Max data had made it very clear that over the period from March to July 1980 the solar constant varied with a period of about 27 days (the solar synodic rotation rate), but Martin Woodward and Hugh Hudson of the University of California were keen to establish whether it would also display a five minute period. Within a year they announced that they had discovered solid evidence for solar irradiance oscillations with various discrete periods of around five minutes and showing amplitudes of about one or two parts per million. These revelations provided new impetus to solar physicists which would enable them to better understand the process occurring inside the Sun.

On the other side of the Pacific Ocean, the Japanese space program had gained momentum during the 1970s and 1980s, and by August 1991 they had launched a solar observatory spacecraft of their very own—Yohkoh (the Japanese word for 'sunbeam'). Weighing just 400 kilograms, Yohkoh was considerably smaller than Solar Max, but its payload of scientific instruments, which was designed a full decade after Solar Max, was far superior in many ways. Yohkoh carried a Hard X-ray Telescope (HXT) operating in the 15 to 100 keV range, a soft X-ray telescope (SXT) covering the X-ray range from 0.2 to 4 keV and visible light from 460 to

480 nm, a Wide Band Spectrometer (WBS) operating in the X-ray and gamma ray wavebands, and a Bragg Crystal Spectrometer (BCS) providing very high spectral resolution in four soft X-ray bands centered on 0.178 nm (the Iron XXVI line), 0.186 (Iron XXV), 0.318 nm (Calcium XIX) and 0.506 nm (Sulfur XV). Clearly, Yohkoh was fully equipped to extend the study of solar flares.

The data garnered by Yohkoh yielded brand new insights into the structure of solar flares demonstrating for the first time that electrons accelerating along magnetic loops into the chromosphere, produce white-light flares. In addition, Satoshi Masuda of the University of Tokyo and colleagues discovered 'bright knots' seen in soft X-rays at the top of the flaring loops as well as bright 'kernels' using Yohkoh's hard X-ray imager located above some of the soft X-ray loops. This came as a great surprise, as up to that time hard X-rays had only been observed at the foot of the loop. In addition, the hard X-ray kernels appeared to coincide with magnetic field reconnection sites, but it was hard to see how such a hard X-ray source could be confined to the apex of a loop for a prolonged period of time. Thus, like all good scientific discoveries, it raised more questions than it had answered (Fig. 9.6).

The soft X-ray telescope on board Yohkoh also imaged numerous small 'microflares', as they came to be known, erupting in magnetic loops around some active areas. These microflares had energies ranging from 10^{22} Watts (equivalent to the energy released during a small solar flare) down to 10^{18} Watts, near the threshold of measurement for the instrument. In all, over one hundred were observed in some active regions, brightening and fading on timescales of a few tens of minutes, yet in other active regions scarcely any microflares were observed. These were found to contribute substantially to the soft X-ray intensity of active regions, although Shimizu showed that they weren't numerous enough to explain the observed coronal heating, the source of which was still unknown.

Prior to the launch of Yohkoh it was widely believed that the Sun loses most of its mass along open coronal hole field lines, because it was thought that the much denser plasmas observed around these active regions were contained by the closed magnetic loops. Although some mass was undoubtedly shed via CMEs, this was still understood to be a relatively insignificant mode of

FIG. **9.6** Artist's impression of the Japanese Yohkoh spacecraft (Image credit: NASA)

mass loss. But after analyzing Yohkoh's soft X-ray images Kazunari Shibata, Y. Uchida and colleagues identified a novel mass loss mechanism. Immediately above some active regions, they discovered long X-ray columns moving outwards from the outer corona at velocities of up to 200 Km/s. The initiating process was thought to be similar to that of microflares, that rely on magnetic reconnection but, significantly, this mass was apparently being ejected along open field lines (Fig. 9.7).

After more than a decade of successful observations, Yohkoh terminated its mission when it was placed in its "safe hold" mode and so could no longer lock onto its target. Unfortunately, this occurred just as an annular eclipse of the Sun began on December 14, 2001. Operational glitches conspired with other flaws in such a way that its solar panels were no longer capable of charging the spacecraft's batteries. That said, a few other solar eclipses had successfully been monitored by the spacecraft over its many years of service.

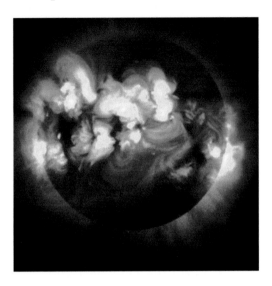

FIG. 9.7 The Sun during an active time, with lots of Solar flares (Image credit: Yohkoh Solar Observatory)

Finally, Yohkoh burned up during its atmospheric re-entry over South Asia on September 12, 2005.

The Sun could only be observed from low solar latitudes as recently as 1990. The ecliptic plane defines the Earth's orbit, which is inclined from the Sun's equatorial plane by only 7.25 degrees. All orbiting solar observatories directly orbiting the Sun did so in planes that were at or close to the ecliptic. The reason for this pertained to the difficulty of a direct launch into a highly inclined solar orbit, which in turn necessitated a prohibitively large launch vehicle. That being the case, several spacecraft including Mariner 10, Pioneer 11, and Voyagers 1 and 2 had already performed gravity assist maneuvers during the 1970s. Those maneuvers were enacted in order to allow them to travel to other planets which also enjoy orbital planes close to—but not exactly in—the ecliptic plane, so these underwent mostly in-plane changes. But gravity assists are not confined to in-plane maneuvers. For example, any planned flyby of Jupiter would normally require a significant plane change. As more and more spacecraft successfully demonstrated this proof of concept an Out-Of-The-Ecliptic mission (OOE) solar mission was soon proposed.

Originally, two spacecraft were to be built by NASA and ESA, called the International Solar Polar Mission. One was designed to fly over Jupiter, then under the Sun and the other vice versa. Both missions would provide simultaneous coverage so that the maximum amount of data could be garnered. Unfortunately, owing to cutbacks, the US mission was mothballed in 1981.

The Ulysses spacecraft was designed and built by ESA. NASA would provide the Radioisotope Thermoelectric Generator (RTG) as well as providing the launch services. The task of fabricating the nine instruments flown on board Ulysses would be split into teams from universities and research institutes across Europe and the United States.

Because of this division of labor, the launch of Ulysses was delayed from February 1983 to May 1986 where it was to be deployed by the Space Shuttle Challenger, but the disastrous explosion of that shuttle mission introduced further delays, which pushed the launch date to October 1990 (Fig. 9.8).

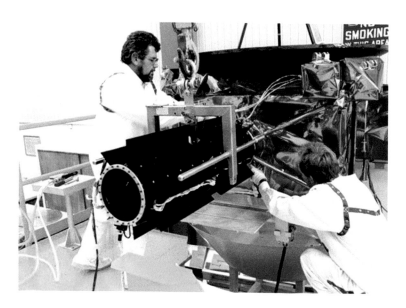

FIG. 9.8 Engineers inspect the spacecraft Ulysses prior to launch (Image credit: NASA)

Ulysses was carried into space by the Space Shuttle Discovery on October 6 1990 and carried five European and four American experiments. The state-of-the-art spacecraft was to investigate the solar wind, the Sun's magnetic field, solar flare emissions, as well as studying the chemistry and physics of the gas and dust making up interplanetary space, as well as galactic cosmic rays. The spin-stabilized spacecraft weighed in at 370 kilos and carried 55 kilograms of experiments. Ulysses communicated with NASA's Deep Space Network on Earth using its 1.6 m diameter parabolic antenna equipped with a tiny 5 watt transmitter. Power for the spacecraft was provided by a RTG generator built by NASA that produced 285 W. Using the RTC generator was not without its problems however, as American environmentalists tried to force NASA to abort the launch of Ulysses on the grounds that the radioactive material in the RTG was potentially lethal in the event of an explosion like the ill-fated Challenger. Fortunately for the Ulysses project scientists, the legal challenge failed.

Ulysses was designed to fly about 430,000 kilometer s from the center of Jupiter on February 9 1992, where it would use the planet's gravity to soar 330 million km above the Sun's south pole in August 1994, where it would reach a solar latitude of about 80 degrees. After that, Ulysses would then cross the ecliptic at a distance of about 175 million kilometers from the Sun in February 1995, and then pass 330 million km above the Sun's north pole by the summer of 1995. Although Ulysses was launched during solar maximum, its first pass over the Sun's poles would coincide with solar minimum.

Many of the interplanetary spacecraft that had flown close to the ecliptic had found that the solar wind velocity was of the order of 350 km/s, although every now and again a faster wind had been detected with a velocity more than twice as large. This 750 km/s component was more frequently observed just before solar minimum, when the polar holes during which time displayed low latitude extensions. In the space of a 13 month period between June 1992 and July 1993, Ulysses found that the solar wind velocity oscillated between a maximum of about 750 and a minimum of 350 km/s but after July the spacecraft only measured the high speed solar wind localized to a hole in the Sun's south polar corona. The wind velocity remained more or less the same as Ulysses

FIG. **9.9** Artist's impression of the Ulysses spacecraft shortly after deployment (Image credit: NASA)

moved from the south pole to the north pole of the Sun, with the notable exception of a two month period between February and March 1995, when the spacecraft crossed the ecliptic and solar equatorial plane going from 20 degrees south to 20 degrees north solar latitude.

Ulysses found that solar winds from the poles have smaller ionized oxygen VII to oxygen VI ratios than the winds sampled at the equator. The explanation for this anomaly was due to the origination of the wind in a region some 400,000 K cooler than the region that produces the slower, equatorial wind. This swifter solar wind was also shown to have a lower ratio of magnesium to oxygen atoms (Fig. 9.9).

After taking measurements of the Oxygen VII to Oxygen VI ion and Mg/O ratios over nine solar rotations at mid solar latitudes, Ulysses showed that the changes in both these ratios occurred more dramatically than that seen with the velocity in the solar wind. Curiously though, both ratios changed rapidly at virtually the same time. Because the Oxygen VII to Oxygen VI ion ratio provides a measure of the temperature of the corona where the solar wind originates, while the Mg/O ratio is indicative of the composition of the chromosphere, Ulysses data unveiled a good correlation between chromospheric composition and the temperature of the overlaying corona, the first instance in which such a correlation had been detected.

The reader will recall that since the 1940s cosmic ray flux detected from the Earth decreased around solar maximum (the so-called Forbush effect). Prior to the launch of Ulysses, astronomers believed that low-energy cosmic rays would be impeded more at low solar latitudes, where the solar magnetic field is more complex owing to the presence of sunspots and differential solar rotation, in comparison with high latitudes, where cosmic rays were being be funneled back toward the Sun via its more radial magnetic field lines. Ulysses was to find no such latitude effect for non-solar cosmic rays. Indeed, they appeared to be equally rejected at all solar latitudes. Furthermore, prior to the launch of Ulysses, it was thought that the Sun's general magnetic field would be stronger near the poles than at the equator. Yet again, Ulysses found that it was quite independent of latitude.

Mission controllers initially planned to terminate the Ulysses mission after the spacecraft had passed once over both solar poles, but Ulysses' instruments were still performing well. So in late 1995 the spacecraft's mission was extended for a second orbit of the Sun, with new polar passes to be conducted in 2000 and 2001. ESA's Science Program Committee approved another extension of the Ulysses mission until March 2009 thereby allowing it to operate over the Sun's poles for a third time during 2007 and 2008.

After it became clear that the spacecraft's RTG would not provide enough power to operate science instruments and keep the attitude control fuel, hydrazine, from freezing, mission controllers were forced to implement instrument power sharing. Up until then, the most important instruments on board the spacecraft had been kept in a constant, active state, whilst the others were simply deactivated when not in use. As Ulysses neared the Sun, its power-hungry heaters were shut down and prioritized instruments kept on. But even this energy-saving strategy could prolong the life of the spacecraft only so much. On February 22, 2008, more than 17 years after its launch, ESA and NASA announced that Ulysses' instrumentation would likely cease within a few months. At another press conference on April 12, 2008 NASA announced that the end date would be July 1, 2008. It was good innings for Ulysses. The spacecraft had operated successfully for over four times its planned lifetime.

ESA's Solar and Heliospheric Observatory (SOHO) was launched into space on December 2 1995 by an American launch vehicle, and was designed to amalgamate the mission objectives of two earlier programs that were cancelled, namely GRIST and DISCO. GRIST, which had been planned in the 1970s for multiple flights on Spacelab, was to provide solar spectroscopy through grazing incidence optics in the extreme ultraviolet. Unfortunately, it was mothballed in 1981 as part of the cutbacks that befell from the ESA-NASA difficulties on the ISPM program. DISCO, on the other hand, was to be a free-flying spacecraft undertaking solar seismology (by measuring the Sun's vibration modes), measurements of solar irradiance, and baseline in-ecliptic solar wind and other measurements in support of the ESA IPM program. It was paramount that DISCO be able to conduct uninterrupted observations of the Sun without interruption and that it was placed beyond the Earth's magnetosphere. It was also important that it had a relatively low radial velocity relative to the Sun to aid in data reduction from the solar seismology package. Mission controllers had planned to position DISCO near to the L1 Lagrangian point between the Sun and the Earth, where the gravitational forces from both bodies balance out at a distance of 1.5 million kilometer s inside the Earth's orbit. However, in March 1983 ESA decided to develop the ISO spacecraft at the expense of DISCO, which was subsequently cancelled.

In a rather curious move, ESA had initiated consultation with officials working on ESA's next scientific space program in June 1982 before they had they had made a final decision whether or not to fly DISCO or ISO as an earlier mission. As a result, plans for an outline SOHO spacecraft mission was submitted to ESA in November 1982, even though the DISCO program was still officially on the table. In the earliest days of planning, SOHO was put forward as an Earth-orbiting solar observatory primarily designed to conduct high-resolution ultraviolet spectroscopy. In early 1983 however, DISCO was cancelled and after that time ESA resolved to include helioseismological and solar wind experiments on SOHO. Like DISCO, it would be placed at the L1 Lagrangian point. In this way, SOHO inherited the main objectives of the cancelled GRIST and DISCO programs.

Japan had launched a small (188 kilograms) solar observatory spacecraft called Hinotori ('Firebird') in 1981, and had ambitions to launch one or two more solar spacecraft over the next decade. Both ESA and NASA also had a number of other solar spacecraft in the pipeline at that time. The most sensible way forward was to organize a collaborative mission between competing space agencies, and so in September 1983 a meeting took place in the USA between NASA, ESA and ISAS (the Japanese Institute for Space and Astronomical Science) officials. The meeting was very productive and a new space mission was agreed, the International Solar-Terrestrial Physics (ISTP) program, where ESA was to contribute both the SOHO and Cluster spacecraft missions, on the assumption that they could both be financed.

Cluster mission was to measure the solar wind at a number of positions simultaneously in near-Earth space by making use of four coordinated spacecraft. The Cluster data would be combined with solar wind data recorded by SOHO. Usually during the planning process ESA would select one spacecraft from a list of possible missions, and that would have resulted in only SOHO or Cluster (at best) being selected. Gerhard Haerendel of the Max Planck institute put forward the idea that the SOHO/Cluster mission be presented to the ESA Member States as a single mission, but that it would involve more funding. This led to detailed discussions between ESA and NASA on a possible collaboration on the SOHO/Cluster program. These discussions were still ongoing in February 1986 when the ESA Science Program Committee approved the joint SOHO/Cluster mission, now called the Solar-Terrestrial Science Program (STSP), as the next ESA science program, provided it could be accomplished within the budgetary constraints laid down by ESA of 400 MAU (1984 economic conditions) (Fig. 9.10).

ESA adopted two distinct strategies to keep costs down. First it persuaded the experimenters to reduce their scientific requirements on the SOHO and Cluster programs. In addition, ESA attempted to persuade NASA to launch SOHO for free on either a space shuttle or another expendable launch vehicle. However,

F<small>IG</small>. **9.10** The SOHO spacecraft being prepared for launch at the SAEF-2 facility of the Kennedy Space Center before being fuelled and encapsulated on top of the Atlas-Centaur AC-121 on pad 36B (Image credit: NASA)

neither of these approaches could meet the budgetary requirements completely, since Cluster required a dedicated Ariane 4 launcher which alone would set them back about 100 MAU. However in 1988, after numerous discussions with the Ariane program, an agreement was forged that would enable Cluster to fly as an experimental payload on one of the Ariane 5 qualification flights, provided ESA funded the additional cost of 13MAU for accommodating Cluster on the launcher. But disaster stuck on June 4 1996 when the first Ariane 5 qualification flight failed less than a minute after lift-off, destroying all four Cluster spacecraft on-board.

In the meantime, the other part of the STSP program, SOHO, had been successfully launched by an American Atlas II-AS on December 2 1995 from Cape Canaveral, and was placed into its halo orbit at the L1 Lagrangian point on February 14 1996, six weeks ahead of schedule. The launch and orbital maneuvers were carried out so well that there was still enough fuel on-board SOHO to keep the spacecraft operational for at least ten years, or about twice as long as was originally envisaged.

The scientific instruments on board SOHO was provided by institutes in 15 different countries, but the launch vehicle was provided by the USA. The payload consisted of three instruments (GOLF, VIRGO and MDI) devoted to helioseismology, five (SUMER, CDS, EIT, UVCS and LASCO) were dedicated to observing the solar atmosphere and corona, and four (SWAN, CELIAS, COSTEP, and ERNE) were designed to monitor the solar wind (Fig. 9.11).

FIG. 9.11 Artist's impression of the SOHO observatory conducting solar observations in space (Image Credit: NASA)

In the early 1980s some scientists expressed their reservations about whether it was a good idea to try helioseismology from space since ground-based measurements were already producing high quality data in this regard. However, after seeing the ACRIM results from Solar Max the scientific community had a change of heart and so the aforementioned instruments were included on the SOHO mission. Within just a few months, SOHO helioseis-mology experiments were returning data of the highest quality. The MDI (Michelson Doppler Imager) instrument, in particular, which was designed to simultaneously measure the vertical motion of the Sun's surface at a million different points every min-ute, produced data never seen before thereby revolutionizing our knowledge of the solar convective zone just below its visible sur-face. These data allowed astronomers to produce the first maps of both horizontal and vertical motions in the Sun's convective layer, revealing vertical convection in more or less evenly spaced columns.

High resolution, visible light studies of the solar photosphere dating back centuries had revealed granules covering the surface of the Sun. These granules, which are usually about 1000 to 1500 km across, tend to be about 30 % brighter than the intergranular lanes that separate one from the other. This corresponded to a tempera-ture differential of about 400 K. Doppler measurements conducted with ground-based solar telescopes indicated that they represent the tops of ascending currents which then move sideways to the intergranular lanes, where the cooler gas then descends. Each gran-ule has longevity of about 20 minutes, with a velocity difference between the upwelling and descending currents of approximately 2 km/s. Scientists were unsure about the depth of these granules however, with estimates varying from about 10 to 100 km.

In the 1950s, solar physicist A.B. Hart discovered a much larger surface structure called supergranulation that should have shown up on the MDI results. That said, supergranulation is a much more subtle structure than granulation and is completely invisible at visual wavelengths, but did show up on spectrohclio-grams which can dctect gaseous motion on the surface of the Sun. Subsequent observations by other astronomers uncovered horizontal gas flows from outwards from the center of these super-granules at a leisurely speed of about 0.4 km/s, with a weak upward flow in the center, and a downward flow near their edges at a velocity of about 0.1 km/s. The supergranules, which last anywhere from 24 to 36 hours, were found to be about 30,000 km across. SOHO's MDI instrument showed, however, that the downward convective motion originated at the center of the supergranules, rather than their periphery.

The Sun's surface was known to pop once every 25 days at the equator and once every 34 days near the poles, but there was still considerable uncertainty over how the interior of the Sun popd. Early models suggested that the Sun's rotation would be constant on cylinders parallel to the Sun's rotation axis. Then in 1988 Kenneth Libbrecht, based at the Big Bear Solar Observatory, discovered that this was not the case using helioseismology, as his data implied that the angular velocity of the Sun is approximately constant along radial lines, for low and medium latitudes, all the way down to the bottom of the convective zone. The SOHO data subsequently confirmed this, showing that below the convective zone the angular velocity is more or less constant and similar to that of the surface at mid-latitudes. This meant that there was an intense shear layer at the base of the convective zone extending beyond solar mid-latitudes. Computer modeling of the shear layer suggested that it would produce a large amount of turbulence and could even be a location where the dynamo that generates the Sun's magnetic field originates.

It was widely anticipated that a Fourier analysis (i.e. frequency) of the solar surface velocities measured by the GOLF instrument, as well as the spectral and total irradiance measurements by VIRGO, would provide evidence of the so-called g-modes (gravity-modulated oscillations) caused by the sloshing of material deep inside the Sun. And while the experiments produced excellent data, with a much higher signal-to-noise ratio than anything achieved prior to that time with Earth-based instruments, no such g-modes could be identified.

SOHO's UVCS (Ultra-Violet Coronograph Spectrometer) instrument took measurements of the velocities of a number of ions in the outer solar corona, to gain a greater understanding of the mechanisms that lead to their acceleration as they move outwards from the Sun. Furthermore, it showed that in coronal holes, proton outflow velocities increased from 50 to 200 km/s between 1.5 and 3.5 solar radii, whilst the line widths of Lyman-alpha increased from 190 to 240 km/s, and those of OVI increased from 100 to 550 km/s. This new comparative data enabled theorists to conclude that a particular type of resonance in MHD waves may well be accelerating the plasma.

Coronal mass ejections (CMEs) were studied in detail by LASCO (Large-Angle Spectroscopic Coronograph), made up from three different coronagraphs which had overlapping fields of view that could monitor the solar corona from 1.1 to 30 solar radii. While a few CMEs were found to have a constant velocity as they left the Sun, many others were constantly accelerated. Some CMEs underwent a sudden acceleration a few solar radii from the Sun. And in a few intriguing cases other CMEs were actually observed on both sides of the Sun simultaneously!

A most fascinating CME was observed by the LASCO C2 coronagraph on January 6 1997. It was apparently directed toward the Earth, and was seen as a 'halo' or 'ring' event surrounding the Sun. About 120 minutes after it was first sighted, the CME had enlarged enough to be visible in the LASCO C3 coronagraph, which was equipped with a larger occulting disk. The halo was observed in the C3 instrument for another four and a half hours before it attenuated away into invisibility, but at 04:45 UT on January 10 both the SOHO and NASA's WIND spacecraft detected the same cloud moving at a sprightly 450 km/s. While these observations were being carried out, NASA's Polar spacecraft also detected the interaction of the cloud with the Earth's Van Allen belts, the intensity of which was found to increase by a factor of more than 100 over the next few hours or so. A day later, the Telstar 401 communications spacecraft malfunctioned, possibly due to these radiation-induced effects. These solar storms have been known to have even more serious effects at high latitudes (nearer the magnetic poles) on the Earth, causing serious problems with the distribution of electricity using power lines. As a result, a series of early warning system became set up, based on data provided spacecraft images, to warn electricity suppliers on the ground in advance and operators of spacecraft in orbit of the possible damage that they could cause (Figs. 9.12 and 9.13).

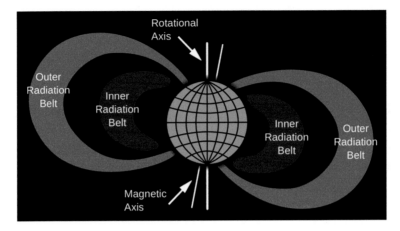

Fig. 9.12 Showing the Earth's Van Allen Belts generated by the planet's magnetic field (Image credit: NASA)

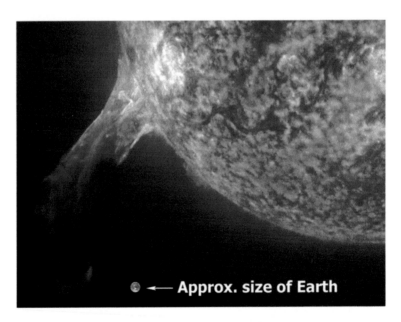

Fig. 9.13 An erupting solar prominence captured by the SHO spacecraft in July 2002 (Image Credit: NASA)

Both the SUMER and CELIAS instruments carried on board SOHO conducted measurements on abundances of many elements in the solar chromosphere and corona to aid in the elucidation of the processes that feed and accelerate the solar wind. These instruments showed an enrichment of up to a factor of some 20 elements with a first ionization energy of less than 10 eV in comparison with their photospheric abundances. Although the partial correlation observed between low values of FIP and high coronal enrichment has been known for some time, the greater sensitivity of SOHO added many more new elements and isotopes to the database, which greatly facilitated theoretical analysis. The high temporal resolution of CELIAS proved indispensable in elucidating how the abundance enhancement changed with structural and velocity variations in the solar wind and corona (Fig. 9.14).

Early results garnered from the high-energy particle detectors (COSTEP and ERNE instruments) monitoring the solar wind on board SOHO were disappointing because the spacecraft was

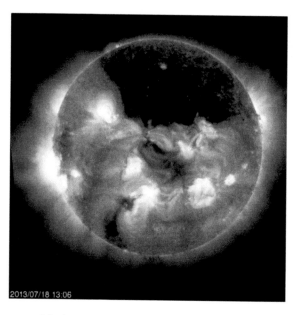

2013/07/18 13:06

Fᴵɢ. 9.14 A coronal hole in the sun was captured by NASA's SOHO observatory. The dark, cool spot of the sun's atmosphere covered a quarter of the sun (Image credit: NASA)

launched during a time of solar minimum. That said, the highest energy instrument carried into space on board SOHO, ERNE, was able to measure and analyze galactic cosmic rays, uncovering significant amounts of hydrogen, helium, boron, carbon, nitrogen, oxygen, neon, magnesium, silicon and iron, the heavier nuclei probably coming from supernovae.

On June 24, 1998, the SOHO team conducted a series of spacecraft gyroscope calibrations and maneuvers. Mission controllers attempted to recover the observatory, but SOHO entered the emergency mode again on June 25. Recovery efforts continued, but all contact with the spacecraft was lost. SOHO was spinning out of control, losing electrical power all the while, and no longer pointing at the Sun.

A team of specialist ESA engineers travelled to the United States to aid in the SOHO recovery operations. Several days went by without contact with the spacecraft. On July 23 1998, staff at the Arecibo Observatory and DSN antennas were employed to locate SOHO with radar, and to pinpoint both its location and attitude. They found that SOHO was quite close to its predicted position, pointing toward the Sun, and was rotating at one RPM. Once SOHO was located, plans for contacting SOHO were formed. On August 3 a message was received from SOHO, the first such signal since the end of June. After days of charging the battery, a successful attempt was made to modulate the carrier and downlink telemetry on August 8. After instrument temperatures were downlinked on August 9, data analysis was performed, and planning for the SOHO recovery began in earnest.

The SOHO Recovery Team began by addressing the limited electrical power. This was followed by the determination of SOHO's anomalous orientation in space. On August 12 the frozen hydrazine fuel tank was thawed out using SOHO's thermal control heaters. Next, the pipes and the thrusters were re-heated to renew the flow of fuel. Then on September 16 SOHO was re-oriented so that it properly faced the Sun again. The following week the SOHO spacecraft resumed routine operations on September 25. Finally, the recovery of the instruments began with SUMER on October 5, and ended on October 24, 1998 with CELIAS.

Despite these recovery operations, only one functioning gyroscope remained and on December 21 1998 it too failed. Mission controllers were forced to fire manual thruster to temporarily regain attitude of the spacecraft. But this procedure consumed 7 kg of fuel every week. In the meantime, ESA developed a new gyroless operations mode that was successfully implemented on February 1, 1999. SOHO continued to provide new insights into our Sun, the likes of which could not have been anticipated when it was first launched back in 1995.

One of the unanticipated spinoffs of the SOHO mission was the discovery of new comets by blocking out the Sun's glare. Indeed one-half of all known comets have been spotted by SOHO's LASCO instrument. Many of these were discovered by ordinary people from 18 different countries around the world searching through the publicly available SOHO images that were placed on the World Wide Web. Michał Kusiak of the Polish Jagiellonian University (Uniwersytet Jagielloński) discovered SOHO's 1999th and 2000th comets on 26 December 2010. And as of April 2014, SOHO has discovered over 2700 comets, with a new comet being uncovered once every three days on average. UK-based amateur, Mike Oates, discovered over 140 comets, many of which were sun grazers. To honor his work, the minor planet "68948 Mikeoates" was named after him (Fig. 9.15).

There have been several other solar observatories launched into space after SOHO. Hinode, formerly known as Solar-B, was a Japan Aerospace Exploration Agency Solar mission in collaboration with the United States and the United Kingdom. It was essentially a follow-up to the Yohkoh (Solar-A) mission. Hinode was launched on the final flight of the M-V-7 rocket from Uchinoura Space Center, Japan on 22 September 2006 at 21:36 UTC. Initial orbit was perigee height 280 km, apogee height 686 km, inclination 98.3 degrees. Then the satellite maneuvered to the quasi-circular sun-synchronous orbit over the day/night terminator, which allows near-continuous observation of the Sun. On 28 October 2006, the probe's instruments captured their first images.

The data from Hinode are being downloaded to the Norwegian, terrestrial Svalsat station, operated by Kongsberg a few kilometers west of Longyearbyen, Svalbard. From there, data are transmitted

242   Space Telescopes

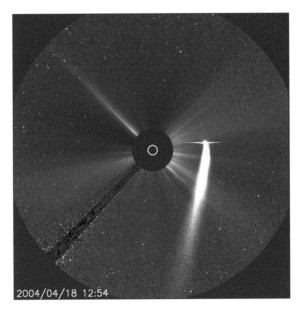

FIG. 9.15 SOHO captures Comet 422 grazing the Sun (shown occulted in center) (Image credit: NASA)

by Telenor through a fiber-optic network to mainland Norway at Harstad, and on to data users in North America, Europe and Japan (Fig. 9.16).

Hinode was planned as a three-year mission to explore the magnetic fields of the Sun. It carried a coordinated suite of optical, extreme ultraviolet (EUV), and x-ray instruments to study the interaction between the Sun's magnetic field and its corona in order to improve our understanding of the mechanisms that power the solar atmosphere and drive solar eruptions. Hinode's EUV imaging spectrometer (EIS) was built by a consortium led by the Mullard Space Science Laboratory (MSSL) in the UK. NASA contributed three science instruments: the Focal Plane Package (FPP), the X-Ray Telescope (XRT), and the Extreme Ultraviolet Imaging Spectrometer (EIS), sharing operations support and managing scientific priorities.

Hinode carries three main instruments to study the Sun: a 0.5 meter Gregorian optical telescope with an angular resolution of about 0.2 arc seconds and covering a field of view measuring 400×400 arc seconds. The Focal Plane Package (FPP) built by the Lockheed Martin Solar and Astrophysics Laboratory in Palo Alto,

FIG. **9.16** Artist's impression of the Hinode spacecraft (then known as Solar-B) in orbit (Image credit: NASA)

California, consists of three optical instruments: the Broadband Filter Imager (BFI) which can image of the solar photosphere and chromosphere in six wide-band interference filters; the Narrowband Filter Imager (NFI) with its adjustable Lyot-type birefringent filter, capable of generating magnetogram and Dopplergram images of the solar surface; and the Spectropolarimeter (SP), which produces more sensitive vector magnetograph maps of the photosphere than any previous solar spacecraft. The FPP also includes a Correlation Tracker (CT) which locks onto solar granulation to stabilize the SOT images to a fraction of an arc second. The spatial resolution of the SOT is represented a five-fold improvement over previous space-based solar telescopes such the MDI instrument carried on SOHO.

The XRT (X-ray Telescope) consisted of a modified Wolter I telescope employing grazing incidence optics to image the solar corona's hottest components (0.5 to 10 million K) with 1 arc-second-sized pixels on the CCD array. This telescope could obtain images over a generous 34 arcminutes and fully capable of imaging the full solar disk. The telescope was designed and built by Smithsonian Astrophysical Observatory (SAO), in collaboration with the Harvard-Smithsonian Center for Astrophysics (CfA). This state-of-the-art camera was developed by NAOJ and JAXA.

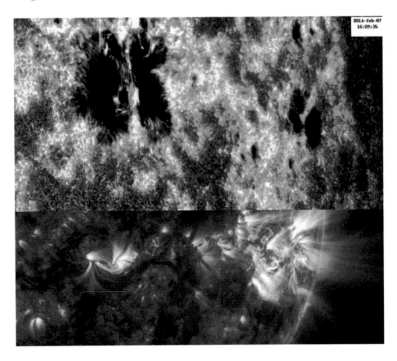

Fig. 9.17 Active Region 11967 consisted of a major sunspot that released numerous solar flares in early 2014. Above is an image taken by the Solar Optical Telescope (SOT) aboard Hinode in the Calcium II H line on Feb. 7, 2014. Below is a wider view of Active Region 11967 as seen by SDO on Feb.12, 2014 (Image credit: NASA)

The EIS (Extreme-Ultraviolet Imaging Spectrometer) is a normal incidence extreme ultraviolet (EUV) spectrometer that captures well resolved spectra in two EUV wavelength bands: 17.0–21.2 and 24.6–29.2 nm. Its spatial resolution is around 2 arc seconds, and the field of view is approximately 560×512 arc seconds. These spectral lines are emitted at temperatures ranging from 50,000 K to 20 million K. EIS is used to identify the physical processes involved in heating the solar corona (Fig. 9.17).

By conducting an ongoing study of the sun's magnetic field, scientists hope to shed new light on explosive solar activity that can interfere with satellite communications and electric power transmission grids on Earth and threaten astronauts on the way to or working in Earth orbit. In particular, scientists wish to determine if they can identify the magnetic field configurations that

Fɪɢ. **9.18** Hinode view of the 2012 Venus transit (Image credit: Hinode Solar Observatory)

lead to such explosive episodes on the Sun and use this information to predict when these events may occur (Fig. 9.18).

This brings to an end to our exploration of the star that nurtures all life on Earth. But it is worth reminding the reader that the Sun is actually the most stable star known. And though many stars bear some superficial resemblance to the Sun, astronomers have yet to identify an identical twin. The Sun's orbit about the Galactic center is very nearly circular, avoiding dangerous interactions with giant clouds and gas and dust that would radically destabilize the solar system and the Earth in particular. The Sun and its retinue of planetary bodies are located a safe distance from the inhospitable parts of the galaxy near its center. Being located on the outskirts of the Milky Way, we enjoy gloriously dark skies that allow us to see far back into the cosmic history, just a few hundred thousand years after the Big Bang. In so doing, we can figure out the events which led to the origin, evolution and fate of our universe. How fortunate we are to be able to do so!

In the next chapter, we shall explore who astronomical satellites have transformed our understanding of the vast distances between the stars and the discovery of extrasolar planetary systems that inhabit the cosmos.

10. Measuring the Heavens

How far the stars? Do they have planets? If so, what are these planetary systems like? Do any of them harbor life? Never before in the history of humankind have we been able to answer these questions in an objective way, but in the last few decades new technologies have enabled us to go well beyond the realm of pure speculation to see for ourselves what the truth is. In this chapter, we'll be taking a closer look at the way mankind has mapped the star-filled heavens and uncovered a great variety of worlds orbiting those stars. Yet, like all other areas of science, these new discoveries were on the crest of a wave of intellectual enquiry harking back to ancient times and reverberating down the centuries.

The Pythagorean school of Greek philosophers are widely regarded by scholars as being the first to suggest that the Earth is spherical, as early as the sixth century BC, but it was Aristotle in his *De Caelo* around 340 BC, who gives mention to the first scientific measurements of the Earth's size by Eratosthenes in about 240 BC. And while the principle of the measurement of the Earth–Moon distance had already been established by Aristarchus of Samos around 250 BC, it was not for another 130 years that Hipparchus first calculated the distance of the Moon from the Earth in 120 BC by measuring the Moon's parallax. Furthermore, a comparison of Hipparchus' star catalogue of 1080 stars with the work of his predecessors led to the discovery of the precession of the equinoxes and the eccentricity of the Sun's path. Remarkably, these advances were made using measurements conducted with the naked eye, the resolution power of which is limited to a just few minutes of arc owing to the primitive sighting instruments they employed.

Like many other areas of scientific knowledge, little advance was made in the science of astrometry during the millennium of the 'Middle Ages', in which western civilization largely stagnated, retaining with the concept of a universe with the Earth at its center. However, the awakening of man's scientific curiosity at the

© Springer International Publishing Switzerland 2017
N. English, *Space Telescopes*, Astronomers' Universe,
DOI 10.1007/978-3-319-27814-8_10

time of the Reformation led to revised interest in astrometry. The Polish astronomer and mathematician Nicolaus Copernicus propounded the heliocentric concept in 1543, and his Danish counterpart, Tycho Brahe, using his brass azimuth quadrant and many other new instruments, carried out a long series of observations during the second half of the sixteenth century. These observations were to provide the basis for the laws of planetary motion as first proposed by Johannes Kepler between 1609 and 1619.

Although observations were still being made with the naked eye, this was soon to change. By 1609, Galileo began to make use of the optical telescope, using information obtained from Holland (where the first instrument was fabricated sometime around in 1604). With it, enormous advances in astrometry were soon to be made. Over the subsequent decades, the angular error in astrometric measurements fell to about 15 seconds of arc by 1700, and to just 8 seconds of arc by 1725. This made it possible to detect stellar aberration—the small positional displacements of the velocity of light to the Earth's orbital velocity—and nutation—a perceptible 'wobbling' in the Earth's spin axis produced by the gravitational influence of the Sun and Moon with a period of 18.6 years.

The rate of precession was worked out by Sir Edmund Halley, who compared contemporary observations with those that Hipparchus and others had made. While most of the stars displayed a general drift amounting to a precession of about 50 seconds of arc per year, Halley announced in 1718 that three stars, Aldebaran, Sirius and Arcturus, had shifted from their expected positions by large fractions of a degree. Halley further deduced that each star had its own 'proper motion'. Eventual improvements in observational precision during the eighteenth century unveiled the motions of many more stars, and in 1783 the Anglo-German astronomer, Sir William Herschel, found that he could partly explain these motions by assuming that the Sun itself was moving. This suggested that some stars might be relatively close to the Sun, and so astronomers redoubled their efforts to detect 'trigonometric parallax', that is, the apparent oscillation in a star's position arising from the Earth's annual motion around the Sun. The German astronomer, Friedrich Bessel, was the first to publish such a parallax value back in 1838, following his studies of the motion of 61 Cygni. Bessel's careful and detailed analysis of the measurement

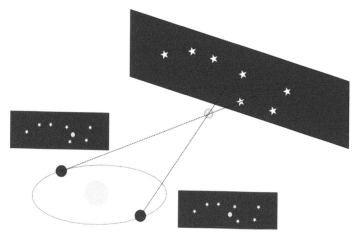

Fɪɢ. **10.1** Schematic diagram showing the principle of trigonometric parallax (Image credit: https://commons.wikimedia.org/wiki/File:ParallaxeV2.png)

errors and his use of both coordinates on the sky gave credibility to his results, after many previous claims from astronomers to have measured a stellar parallax. Thomas Henderson is credited with the first measurement of stellar parallax from the southern hemisphere, when he conducting measurements on the bright star Alpha Centauri at the Cape of Good Hope, in 1832–33, though admittedly he did not analyze the measurements for some years. We now understand that the two components of this star, together with a faint companion called Proxima Centauri, form the nearest known star system to the Sun, at a distance of a little more than 4 light years. And in 1837 to 1838, yet another German astronomer, Wilhelm Struve, measured the parallax of Vega in similar fashion (Fig. 10.1).

Observations improved substantially with the application of photography to astronomical science, which gave astrometry yet another tool it could exploit. In 1887 a world-wide cooperative program was started to conduct a full photographic survey of the sky. Eighteen countries were involved in this project, collectively known as the 'Carte du Ciel'. All observatories involved agreed to employ the same design of astrograph, plates, and observing protocols. Together, these produced a precision of around 1 arc second for a grand total of 13 million stars. Since the early part of twentieth century determinations of photographic trigonometric parallaxes have been made at more than a dozen observatories. The technique involved measuring the shift of selected stars relative to a few stars in the same field of view over a number of years. Several thousand trigonometric parallaxes had now been measured from the ground; however, only a few hundred were considered to be known with an accuracy of better than about 20 %, while the systematic effects remained conspicuous but uncertain. By the second half of the twentieth century, the accurate measurement of star positions from the ground was running into essentially insurmountable barriers to improvements in accuracy, especially for large-angle measurements and systematic terms. Problems were dominated by the effects of the Earth's atmosphere, but were compounded by complex optical terms, thermal and gravitational instrument flexures, and the absence of all-sky visibility. A formal proposal to make these exacting observations from space was first put forward in 1967.

Although originally proposed to the French space agency CNES, it was considered too complex and expensive for a single national program. Its acceptance within the European Space Agency's scientific program, in 1980, was the result of a lengthy process of study and lobbying. That being the case, there were good reasons why more precise positional measurements ought to be made in order to answer fundamentally new questions in astronomy and astrophysics. The observational situation promised to make a rapid and dramatic change when, at the behest of the scientific community, and following an internal feasibility study supported by its member state scientists, the European Space Agency (ESA) embarked upon project Hipparcos, the mission of which was to conduct precise positional measurement of some 120,000 stars in the solar neighborhood. Entering orbit astride an Ariane 4 rocket on August 8 1989, Hipparcos was to produce two major catalogs—called Hipparcos and Tycho—of star positions, parallaxes, and proper motions, along with photometric and other data on the stars observed.

The underlying scientific motivation behind Hipparcos was to determine the physical properties of the stars through the measurement of their distances and space motions, and thereby place theoretical studies of stellar structure and evolution, and studies of galactic structure and kinematics, on a more secure empirical basis. Observationally, the spacecraft's objective was to provide the positions, parallaxes, and annual proper motions for some 100,000 stars with an unprecedented accuracy of 0.002 seconds of arc. The name of the space telescope, "Hipparcos" was an acronym for High Precision Parallax Collecting Satellite, and it also reflected the name of the ancient Greek astronomer Hipparchus, who is considered the founder of trigonometry and the discoverer of the precession of the equinoxes owing to the Earth's precessional motion.

The satellite was operated by the ESA operations control center at ESOC, Darmstadt, Germany. Including an estimate for the scientific activities related to the satellite observations and data processing, Hipparcos mission cost about €600 million (2000 economic conditions), and its execution involved some 200 European scientists and more than 2000 individuals in European industry.

Remarkably, Hipparcos carried a single all-reflective, eccentric Schmidt telescope, with an aperture of just 29 cm (11.4 in). A special beam-combining mirror allowed two fields to be superimposed and 58 degrees apart, into the common focal plane. This complex optical system consisted of two mirrors tilted in opposite directions, each occupying half of the rectangular entrance pupil, providing an unvignetted field of view of about 1 square degree. The telescope used a system of grids located at the focal surface, composed of 2688 alternate opaque and transparent bands, with a period of 1.208 arcseconds. Behind this grid system, an image dissector tube (photomultiplier type detector) was fitted with a sensitive field of view of about 38-arc-second diameter, and which converted the incoming light into a sequence of photon counters that allowed the phase of the entire pulse train from a star could be derived. The apparent angle between two stars in the combined fields of view was then obtained from the phase difference of the two star pulse trains. Although originally planned to observe some 100,000 stars with an astrometric accuracy of about 0.002 arc seconds, the final Hipparcos Catalog comprised nearly 120,000 stellar sources with a median accuracy of slightly better than 0.001 arc seconds (1 milli-arcsecond).

An additional photomultiplier system intercepted a beam splitter in the optical path and was used as a means of mapping a star, as well as to monitor and determine the satellite attitude, and in so doing gathering photometric and astrometric data of all stars down to the 11th magnitude. These measurements were made in two broad bands approximately corresponding to B and V in the traditional Johnson UBV photometric system. The positions of these latter stars were to be determined to a precision of 0.03 arc-second, which was 25 times less precise than the main mission stars. Originally targeting around 400,000 stars, the resulting Tycho Catalog comprised just over 1 million stars, with a subsequent analysis extending this to the Tycho-2 Catalog of about 2.5 million stars.

The attitude of the Hipparcos satellite was controlled to scan the celestial sphere in a regular precessional motion, maintaining a constant inclination between the spin axis and the direction to the Sun. The spacecraft spun around its Z-axis at a rate of 11.25 revolutions/day (168.75 arcsecond per second) at an angle of 43° to the Sun. In turn, the Z-axis of the spacecraft rotated about the sun-satellite line at 6.4 revolutions per year (Fig. 10.2).

The spacecraft consisted of two platforms and six vertical panels, fabricated from aluminum honeycomb. Its solar array consisted of three deployable sections, generating around 300 W in

FIG. 10.2 European Space Agency's Hipparcos star-survey satellite (Image credit: ESA)

total. Two S-band antennae were located on the top and bottom of the spacecraft, providing an omni-directional downlink data rate of 24 kbit/s. An attitude and orbit-control subsystem (comprising 5 Newton hydrazine thrusters for course maneuvers, 20 mili-Newton cold gas thrusters for attitude control, and gyroscopes for attitude determination) ensured that the spacecraft correct dynamic attitude control during the operational lifetime of 3.5 years.

It is no exaggeration to say that the data from Hipparcos had a profound influence on most fields in astronomy and astrophysics. The improved astrometric precision led to much better estimates of many of the basic properties of stars, from their luminosities to their chemical composition. This alone significantly advanced our understanding of the internal constitution of stars and their evolution. With a robust and precise reference frame, astronomers could finally describe the dynamics of stars in the solar neighborhood and study many stellar clusters in unprecedented detail.

Beyond our own galaxy, the Milky Way, stellar distances based on Hipparcos's parallax measurements allowed cosmologists to refine the calibration of the cosmic distance ladder, leading to a more precise estimate of the expansion rate of the universe and of its age. The new value for the age of the universe (derived from the reciprocal of the Hubble constant discussed in Chap. 1) solved a long-standing conundrum. For the first time in human history, astronomers could show that the universe was older than the oldest globular clusters in the Galaxy, which also had their ages revised based on Hipparcos data.

Data from Hipparcos were also applied to the study of exoplanets—something that could not have been foreseen at the time of the mission planning given that the first planet outside our Solar System was found only in 1995. Astronomers have used data from the Hipparcos catalogue to obtain upper limits for the masses of several exoplanets, confirming their nature, and to determine their masses and to characterize the properties of their parent stars. The application of Hipparcos data to this field pointed out that a next-generation, space-based astrometry mission with precision of 1 millionth of an arc second could bring a significant contribution to the study of planetary systems far beyond our own.

Hipparcos data were used to compile the Millennium Star Atlas, a three-volume publication with 1548 charts of the heavens, which was released to the public in 1997. Planetarium software as well as other visualizations of the sky, including Google Sky and the astronomy apps available for smartphones, was also based on data from the Hipparcos mission. The Hipparcos and Tycho 2 catalogues are still routinely employed as references for ground-based telescopes to find their targets, and for the navigation of space missions.

As successful as the Hipparcos satellite was for greatly increasing the accuracy with which humankind measured the distances to the stars, the advent of more sophisticated technology allowed us to improve on this still more.

The Gaia space telescope has its roots in ESA's Hipparcos mission (1989–1993). Its mission was proposed in October 1993 by Lennart Lindegren (Lund University, Sweden) and Michael Perryman (ESA) in response to a call for proposals for ESA's Horizon Plus long-term scientific program. It was adopted by ESA's Science Programme Committee as cornerstone mission number 6 on 13 October 2000, and the B2 phase of the project was authorized on 9 February 2006, with EADS Astrium taking responsibility for the hardware. The name "Gaia" was originally derived as an acronym for Global Astrometric Interferometer for Astrophysics. This reflected the optical technique of interferometry that was originally planned for use on the spacecraft. However, the working method has since changed, and although the acronym is no longer applicable, the name Gaia was retained to provide continuity with the project (Fig. 10.3).

The total expenditure of the mission came around €740 million (~$1 billion), including the cost of manufacture, launch and ground operations. In the end though, Gaia was completed two years behind schedule and at 16 % above its initial budget, owing to difficulties encountered in polishing Gaia's ten mirrors and assembling and testing the focal plane camera system, as well as a number of other minor glitches.

Gaia was launched by Arianespace, using a Soyuz ST-B rocket with a Fregat-MT upper stage, from the Ensemble de Lancement Soyouz at Kourou in French Guiana on the morning of 19 December 2013. The satellite separated from the rocket's upper stage just

Fig. 10.3 Artist's impression of the Gaia spacecraft in orbit (Image credit: ESA)

43 minutes after launch and immediately headed towards the Sun–Earth Lagrangian point L2 located approximately 1.5 million kilometers from Earth, arriving there on January 8 2014. The L2 point provides the spacecraft with a very stable gravitational and thermal environment. There it settled into a so-called Lissajous orbit, which avoided blockage of the Sun by the Earth, and which would limit the amount of solar energy the satellite could produce through its solar panels, as well as disturbing Gaia's thermal stability. Shortly after Gaia's launch, a 10-metre diameter sunshade was deployed. The sunshade always faces the Sun, thus cooling all the spacecraft's telescopic components and powering Gaia using solar panels on its surface.

The Gaia payload consists of three main instruments:

The astrometry instrument (Astro), which was designed to pinpoint the positions of stars of magnitude 5.7 to 20 by measuring their angular motion. By combining the measurements of any given star over the five-year mission, Gaia will be able to determine its parallax, and therefore its distance and proper motion (the velocity of the star as it moves on the plane of the sky).

The photometric instrument (BP/RP) allowed Gaia to acquire luminosity measurements of stars over the 320–1000 nm spectral band, and over the same magnitude 5.7–20 range. The blue and red photometer s (BP/RP) are used to determine stellar properties such as temperature, mass, age and elemental composition. Multi-color photometry is provided by two low-resolution fused-silica prisms dispersing all the light entering the field of view in the along-scan direction prior to detection. The Blue Photometer (BP) operates in the wavelength range 330–680 nm; the Red Photometer (RP) covers the wavelength range 640–1050 nm.

The Radial-Velocity Spectrometer (RVS) is used to determine the velocity of celestial objects along the line of sight by acquiring high-resolution spectra in the spectral band 847–874 nm (field lines of calcium ion) for objects up to magnitude 17. Radial velocities are measured with a precision between 1 km/s and 30 km/s. The measurements of radial velocities are important to correct for perspective acceleration which is induced by the motion along the line of sight. The RVS reveals the velocity of the star along the line of sight of Gaia by measuring the Doppler shift of absorption lines in a high-resolution spectrum.

In order to maintain the fine pointing to focus on stars many light years away, there are almost no moving parts. The spacecraft subsystems are mounted on a rigid silicon carbide frame, which provides a stable structure that will not expand or contract due to heat. Attitude control is provided by small cold gas thrusters that can output 1.5 micrograms of nitrogen per second. The telemetric link with the satellite is about 3 Mbit/s on average, while the total content of the focal plane represents several gigabit/s. Therefore only a few dozen pixels around each object can be downlinked.

The Gaia mission aims to construct a 3D space catalog of approximately 1 billion astronomical objects, mainly stars (approximately 1 % of the Milky Way population) brighter than 20G magnitudes, where G represents the Gaia magnitude passband range between 400 and 1000 nanometers light wavelengths. The spacecraft will monitor each of its target stars about 70 times over a period of five years. What is more, Gaia is expected to detect a plethora of Jupiter-sized planets beyond the Solar System, a half a million quasars and many tens of thousands of novel asteroids and comets orbiting the Sun.

Despite its name, it will come as some surprise that Gaia does not actually use interferometry to determine the positions of stars. When the spacecraft was originally designed, interferometry seemed the best way to achieve the target resolution, but the prototype design was later transformed into an imaging telescope. Similar to its predecessor Hipparcos, Gaia consists of two telescopes providing two observing directions with a constant wide angle (106.5°) separating them. The spacecraft rotates continuously around an axis set at right angles to the two telescopes' lines of sight. In addition, Gaia's spin axis undergoes a slight precession across the sky, whilst maintaining the same angle to the Sun. By precisely measuring the relative positions of objects from both observing directions, a robust system of reference is obtained. Both telescopes on board Gaia have 1.45×0.5 m primary mirror as well as a 1.0×0.5 m focal plane array on which light from both telescopes is projected. These consist of 106 CCDs of 4500×1966 pixels each, or a total of one gigapixel.

Each celestial object will be observed on average about 70 times during the mission, which is expected to last until 2018. These measurements will help determine the astrometric parameter s of stars: two corresponding to the angular position of a given star on the sky, two for the derivatives of the star's position over time (motion) and lastly, the star's parallax from which distance can be calculated. The radial velocity of the brighter stars is measured by an integrated spectrometer observing the Doppler Effect. Because of the physical constraints imposed by the Soyuz spacecraft, Gaia's focal arrays could not be equipped with optimal radiation shielding, and ESA expects their performance to suffer somewhat toward the end of the five year mission. Ground tests of the CCDs while they were subjected to radiation provided reassurance that the primary mission's objectives can be met.

The expected accuracies of the final catalogue data have been calculated following in-orbit testing, taking into account the issues of stray light, degradation of the optics, and the basic angle instability. The best accuracies for parallax, position, and proper motion are obtained for the brighter observed stars, apparent magnitudes 3–12. The standard deviation for these stars is expected to be 6.7 micro-arc seconds or better. For fainter stars, error levels increase, reaching 26.6 micro-arc seconds error in the parallax for 15th magnitude stars, and several hundred micro-arc seconds for 20th magnitude stars. For comparison, the best parallax error levels from the new Hipparcos reduction are no better than 100 micro-arc seconds, with typical levels several times larger.

The overall data volume that will be retrieved from the spacecraft during the five-year mission assuming a compressed data rate of 1 Mbit/s is approximately 60 terra bytes (TB), amounting to about 200 TB of usable uncompressed data on the ground, stored in the InterSystems Caché database. The responsibility of the data processing, partly funded by ESA, has been entrusted to a European consortium (the Data Processing and Analysis Consortium, or DPAC) which was selected after its proposal to the ESA Announcement of Opportunity released in November 2006. The funding for the DPAC was provided by the participating nations and has been secured until the production of Gaia's final catalogue scheduled for 2020.

Gaia will send back data for about eight hours every day at about 5 Mbit/s. ESA's two most sensitive ground stations, the 35 m diameter radio dishes in Cebreros, Spain, and New Norcia, Australia, will receive the raw data from the spacecraft.

Gaia will create a precise three-dimensional map of astronomical objects throughout the Milky Way and map their motions, which encode the origin and subsequent evolution of the Milky Way. The spectrophotometric measurements will provide the detailed physical properties of all stars observed, characterizing their luminosity, effective temperature, gravity and elemental composition. This massive stellar census will provide the basic observational data to tackle a wide range of important questions related to the origin, structure, and evolutionary history of our galaxy.

Take a look at the image captured by Gaia above. As the space observatory scans the sky, it measures the positions and velocities of a billion stars with unprecedented accuracy. In addition, for some stars Gaia also determines their speed across the camera's sensor. This data is used in real time by the attitude and orbit control system to ensure the satellite's orientation is maintained within the desired precision.

These velocity statistics are routinely sent to Earth, along with the scientific information, in the form of housekeeping data. They include the total number of stars that are detected every second in each of Gaia's fields of view. These data provide an indication of the density of stars across the sky—that was used to produce this uncommon visualization of the celestial sphere. Brighter regions indicate higher concentrations of stars, while darker regions correspond to patches of the sky where fewer stars are observed.

The plane of the Milky Way, where most of our galaxy's stars reside constitutes the brightest portion of this image, running horizontally and especially bright at the center. Darker regions across this broad strip of stars, known as the Galactic Plane, correspond to dense, interstellar clouds of gas and dust that absorb starlight along the line of sight.

The Galactic Plane is the projection on the sky of the Galactic disc, a flattened structure with a diameter of about 100 000 light-years but with a vertical height of just 1000 light-years. Beyond the plane, only a few objects are visible, most notably the Large and Small Magellanic Clouds, two dwarf galaxies orbiting the Milky Way, that stand out in the lower right part of the image.

A few globular clusters—large assemblies up to millions of stars held together by their mutual gravity—are also sprinkled around the Galactic Plane. Globular clusters, the oldest population of stars in the Galaxy, sit mainly in a spherical halo extending up to 100 000 light-years from the center of the Milky Way.

The globular cluster NGC 104 is also easily visible in the same image (Fig. 10.4), to the immediate left of the Small Magellanic Cloud. Interestingly, the majority of bright stars that are visible to the naked eye and which form the familiar constellations of our night sky are not seen in this image because they are too bright to be used by Gaia's control system. For the same reason, the Andromeda galaxy—the largest galactic neighbor of the Milky Way—also does not stand out here.

Counter intuitively, while Gaia carries a billion-pixel camera, it is not a mission aimed at imaging the sky: it is making the largest, most precise 3D map of our Galaxy, providing a crucial tool for studying the formation and evolution of the Milky Way.

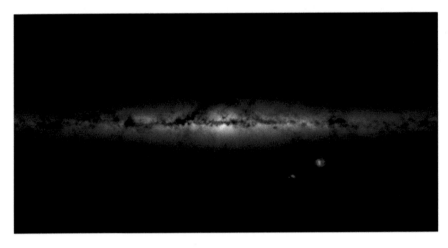

F<small>IG</small>. **10.4** Star density map of the sky, made using Gaia housekeeping data (Image credit: ESA)

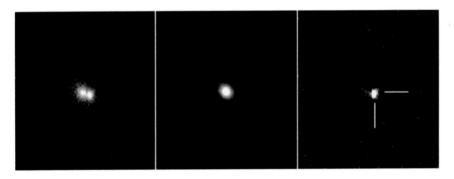

F<small>IG</small>. **10.5** Gaia discovers its first Supernova Gaia14aaa and its host galaxy (Image credit: ESA)

In September 2014, Gaia captured some rare footage of extreme cosmic violence (Fig. 10.5). This powerful event, now named Gaia14aaa, took place in a distant galaxy some 500 million light-years away, and was revealed via a sudden rise in the galaxy's brightness as captured by two Gaia observations separated by one month of time.

Other powerful cosmic events may resemble a supernova in a distant galaxy, such as outbursts caused by the insatiable appetite of the supermassive black hole residing at the galaxy center.

However, in Gaia14aaa, the position of the bright spot of light was slightly offset from the galaxy's core, suggesting that it was unlikely to be associated with a central black hole. Thus, the astronomers looked for more information in the light of this new source. Besides recording the position and brightness of stars and galaxies, Gaia also captures spectra of the objects it images. In fact, Gaia uses two prisms covering the red and blue wavelength regions to produce a low-resolution spectrum that enables astronomers to seek signatures of the various chemical elements present in the source of that light.

As expected in a supernova explosion, the blue part of the spectrum appears significantly brighter than the red part. The best explanation proffered by astronomers is that it might be a 'Type Ia' supernova—the explosion of a white dwarf locked in a binary system with a companion star. While other types of supernovae are characterized by the explosive death throes of giant stars, several times more massive than the Sun.

As we have previously learned, low-mass stars like the Sun end their lives more gently, by puffing up their outer layers and leaving behind a compact white dwarf. Their high density means that white dwarfs can exert an intense gravitational pull on a nearby companion star, accreting mass from it until the white dwarf reaches a critical mass that then sparks a violent explosion. To confirm the nature of this supernova, the astronomers complemented the Gaia data with more observations from the ground, using the Isaac Newton Telescope (INT) and the robotic Liverpool Telescope on La Palma, in the Canary Islands, Spain.

A high-resolution spectrum, obtained on 3 September 2014 with the INT, confirmed not only that the explosion corresponds to a Type Ia supernova, but also provided an estimate of its distance. This provided robust evidence that the supernova happened in the galaxy where it was observed. Supernovae are rare events: only a couple of these explosions happen every century in a typical galaxy. The fact that Gaia captured such an event is all the more remarkable!

At the time of writing, astronomers in the Science Alert Team are still sifting through the data, testing and optimizing their detection software. In the years ahead, they fully anticipate Gaia to discover an additional three new supernovae in distant galaxies every day. In addition to supernovae, Gaia will discover thousands of transient sources of other kinds—stellar explosions on smaller scale than supernovas, flares from young stars coming to life, outbursts caused by black holes that disrupt and devour a nearby star, and quite possibly a suite of entirely new phenomena never before observed.

The Kepler Space Telescope

The quest to understand the likely number and variety of planets orbiting other stars took a huge leap forward with the launch of the Kepler astronomical satellite. Kepler is a space observatory launched by NASA to discover Earth-like planets orbiting other stars. The spacecraft, named after the German Renaissance astronomer, Johannes Kepler (1571–1630), was launched in January 2006. The satellite's launch was delayed eight months because of budget cuts and consolidation at NASA. It was delayed again by another four months in March 2006 due to funding issues. At this time, the high-gain antenna was changed from a gimbal-led design to one that was attached to the frame of the spacecraft in order to reduce expenses and complexity, but at the cost of one observation day per month.

The Kepler observatory was finally launched on March 7, 2009 astride a Delta II rocket from Cape Canaveral Air Force Station, Florida. The launch was a success and all three stages were completed. The cover of the telescope was jettisoned on April 7, 2009, and the first light images were taken on the next day.

On April 20, 2009, the Kepler science team had concluded that further refinement of the focus would dramatically increase the scientific return. On April 23, 2009, they further announced that the focus had been successfully optimized by moving the primary mirror 40 micrometer s (1.6 thousandths of an inch) towards the focal plane and tilting the primary mirror a mere 0.0072 degree.

The spacecraft has a mass of 1039 kilograms (2291 lb) and contains a 1.4-meter (55 in) primary mirror feeding an aperture of 0.95-meter (37.4 in)—at the time of its launch this was the largest mirror on any telescope outside Earth orbit. Manufactured by glassmaker, Corning, using ultra-low expansion (ULE) glass, the mirror was specifically designed to have a mass only 14 % that of a solid mirror of the same size. In order to produce a space telescope system with sufficient sensitivity to detect relatively small planets as they pass in front of their parent stars, a very high reflectance coating on the primary mirror was required. Using ion assisted evaporation, Surface Optics Corp, applied a protective 9-layer silver coating to enhance reflection and a dielectric interference coating to minimize the formation of color centers and atmospheric moisture absorption. Kepler has a field of view of about 12-degrees, roughly equivalent to the size of one's fist held at arm's length. The photometer has a soft focus to provide improved photometric sensitivity, rather than sharp images. An Earth-like transit produces a brightness change of 84 ppm and typically lasts for approximately thirteen hours when it crosses the center of the star.

The focal plane of the spacecraft's camera is made up of 42 CCDs at 2200x1024 pixels, which made it at the time the largest camera yet launched into space, possessing a total resolution of 95 megapixels. The array is cooled by heat pipes connected to an external radiator. The CCDs are read out every six seconds (to limit saturation) and co-added on board for 58.89 seconds for short cadence targets, and 1765.5 seconds (29.4 minutes) for long cadence targets. Due to the larger bandwidth requirements for the former, these were limited in number to 512 compared to 170,000 for long cadence. However, even though at launch Kepler had the highest data rate transmission of any NASA mission, the 29-minute sums of all 95 million pixels constituted more data than could be stored and sent back to Earth. Therefore the science team has pre-selected the relevant pixels associated with each star of interest, amounting to about 6 % of the pixels (5.4 megapixels). The data from these pixels was then re-quantized, compressed and stored, along with other auxiliary data, in the on-board 16 gigabyte solid-state recorder. Data that was stored and downlinked included science stars, p-mode stars, smear, black level, background and full field-of-view images.

On May 13 2009 Kepler successfully completed its commissioning phase and began its search for planets around other stars. On June 19, 2009, the spacecraft successfully sent its first science data to Earth. It was discovered that Kepler had entered safe mode on June 15. A second safe mode event occurred on July 2. In both cases the event was triggered by a processor reset. The spacecraft resumed normal operation on July 3 and the science data that had been collected since June 19 was downlinked that day. On October 14, 2009, the cause of these events was determined to be the malfunctioning of a low voltage power supply that provided power to the RAD750 processor. On January 12, 2010, one portion of the focal plane transmitted anomalous data, suggesting a problem with focal plane MOD-3 module, covering two out of Kepler's 42 CCDs. As of October 2010, the module was described as "failed", but the coverage still exceeded the science goals. NASA has characterized Kepler's orbit as "Earth-trailing". With an orbital period of 372.5 days, the spacecraft slowly lags further behind Earth.

Kepler downlinked roughly twelve gigabytes of data about once per month. On July 14, 2012, one of the four reaction wheels used for fine pointing of the spacecraft failed. While Kepler requires only three reaction wheels to accurately aim the telescope, another failure would leave the spacecraft unable to continue in its mission. This posed a real threat of jeopardizing the mission. On January 17, 2013, NASA announced that one of the three remaining reaction wheels showed increased friction, and that Kepler would discontinue operation for ten days as a possible way of solving the problem. If this second wheel should also fail, the Kepler mission would be over. On January 29, 2013, NASA reported the successful return to normal science collection mode, though the reaction wheel still exhibited elevated and erratic friction levels.

On May 11, 2013, another reaction wheel failed, and the spacecraft was put in point rest state (PRS) on May 15, 2013. In PRS, the spacecraft used a combination of thrusters and solar pressure to control the pointing of the spacecraft. The fuel use was minimized though, which allowed time to attempt recovery of the spacecraft and was automatically placed into a thruster-controlled safe mode with the solar panels facing the Sun and with an intermittent communication link with the Earth. In this state, the fuel would last for several months. Commands were sent to the spacecraft to put it into the so-called Point Rest State. This state further reduced fuel consumption—enough to enable the Kepler probe to work for several years. This state also enabled communication with the spacecraft at any time. In the meantime, work was initiated to get one reaction wheel working again.

In July 2013, the spacecraft remained in point rest state while recovery efforts were planned. By August 15, 2013, attempts to resolve issues with two of the four reaction wheels failed. An engineering report was ordered to assess the spacecraft's remaining capabilities on March 7, 2009.

Designed to survey a portion of our region of the Milky Way to discover Earth-size extrasolar planets in or near the habitable zone and estimate how many of the billions of stars in our galaxy have such planets, Kepler's sole instrument is a photometer that monitors the brightness of over 145,000 main sequence stars without blinking and within a fixed field of view. The data is transmitted to Earth, and then analyzed to detect periodic dimming caused by extrasolar planets that cross in front of their host star.

Kepler is part of NASA's Discovery Program of relatively low-cost, focused primary science missions. The telescope's construction and initial operation were managed by NASA's Jet Propulsion Laboratory, with Ball Aerospace responsible for developing the Kepler flight system. The Ames Research Center was responsible for the ground system development, mission operations since December 2009, and scientific data analysis. The initial planned lifetime was 3.5 years, but greater-than-expected noise in the data, from the stars and the spacecraft, meant additional time was needed to fulfill all of Kepler's mission goals. Initially, in 2012, the mission was expected to last until 2016, but this would only have been possible if all remaining reaction wheels used for pointing the spacecraft remained reliable. On May 11, 2013, a second of four reaction wheels failed, disabling the collection of science data and threatening the future of the mission.

On August 15, 2013, NASA announced that they had given up trying to fix the two failed reaction wheels on board Kepler. This meant the original mission goals needed to be modified, but it did not necessarily mean the end of planet-hunting. NASA had asked the space science community to propose alternative mission plans which would potentially include exoplanet searches, using the remaining two good reaction wheels and thrusters. On November 18, 2013, the K2 'Second Light' proposal was reported. This would include utilizing the disabled Kepler in a way that could detect habitable planets around smaller, dimmer red dwarfs. On May 16, 2014, NASA announced the approval of the K2 extension.

As of January 2016, Kepler and its follow-up observations had found over 2000 confirmed exoplanets in about 500 stellar systems, along with a further 4000 unconfirmed planet candidates. Four planets have been confirmed through Kepler's K2 mission. In November 2013, astronomers reported, based on Kepler space mission data, that there could be as many as 40 billion Earth-sized planets orbiting in the habitable zones of Sun-like stars and red dwarfs within the Milky Way. It is estimated that 11 billion of these planets may be orbiting Sun-like stars. The nearest such planet may be 3.7 parsecs (12 light years) away, according to mission scientists. However, only four of the newly confirmed exoplanets were found to orbit within habitable zones of their related stars: three of the four, Kepler-438b, Kepler-442b and Kepler-452b, are near-Earth-size and likely rocky; the fourth, Kepler-440b, is a super-Earth (Fig. 10.6).

In terms of photometric performance, Kepler is working well, much better than any Earth-bound telescope, but still short of the design goals. The objective was a combined differential photometric

FIG. 10.6 Schematic diagram showing the anatomy of the Kepler spacecraft (Image credit: NASA)

precision (CDPP) of 20 parts per million (PPM) on a magnitude 12 star for a 6.5-hour integration. This estimate was developed allowing 10 ppm for stellar variability, roughly the value for the Sun. The obtained accuracy for this observation has a wide range, depending on the star and position on the focal plane, with a median of 29 ppm. Most of the additional noise appears to be due to a larger-than-expected variability in the stars themselves (19.5 ppm as opposed to the assumed 10.0 ppm), with the rest due to instrumental noise sources slightly larger than predicted. Work is ongoing to better understand, and perhaps calibrate out, instrument noise.

Since the signal from an Earth-size planet is so close to the noise level (only 80 ppm), the increased noise means each individual transit is only a 2.7 σ event, instead of the intended 4 σ. This, in turn, means more transits must be observed to be sure of detection. Scientific estimates indicated that a mission lasting 7 to 8 years, as opposed to the originally planned 3.5 years, would be needed to find all transiting Earth-sized planets. On April 4, 2012, the Kepler mission was approved for extension through the fiscal year 2016, but this also depended on all remaining reaction wheels staying healthy, which turned out not to be the case.

Kepler orbits the Sun avoiding Earth occultations, stray light, as well as gravitational perturbations and torques experienced in a typical Earth orbit. The photometer points to a field in the northern constellations of Cygnus, Lyra and Draco, which is well out of the ecliptic plane, so that sunlight never enters the photometer as the spacecraft orbits. This is also the direction of the Solar System's motion around the center of the galaxy. Thus, the stars which Kepler observes are roughly the same distance from the galactic center as the Solar System, and also close to the galactic plane. This fact is important if position in the galaxy is related to habitability, as suggested by the Rare Earth hypothesis.

Most of the extrasolar planets previously detected by other projects were giant planets, mostly the size of Jupiter and bigger. Kepler is designed to look for planets 30 to 600 times less massive, closer to the order of Earth's mass (for comparison Jupiter is 318 times more massive than Earth). The method used, the transit method, involves observing repeated transit of planets in front of their stars, which causes a slight reduction in the star's apparent magnitude, on the order of 0.01 % for an Earth-size planet. The degree of this reduction in brightness can be used to deduce the diameter of the planet, and the interval between transits can be used to deduce the planet's orbital period, from which estimates of its orbital semi-major axis (using Kepler's laws) and its temperature (using models of stellar radiation) can be calculated.

The probability of a random planetary orbit being along the line-of-sight to a star is the diameter of the star divided by the diameter of the orbit. For an Earth-like planet at 1 AU transiting a Sun-like star the probability is only about 0.47 % or about 1 chance in 210. For a planet like Venus orbiting a solar type star the probability is slightly higher, at 0.65 %; such planets could be Earth-like if the host star is a late G-type star such as tau Ceti. If the host star has multiple planets, the probability of additional detections is higher than the probability of initial detection, assuming planets in a given system tend to orbit in similar planes—an assumption consistent with current models of planetary system formation. For instance, if a Kepler-like mission conducted by aliens observed Earth transiting the Sun, there is a 12 % chance that it would also see a 'Venus' transiting (Fig. 10.7).

Fig. 10.7 Artist's impression of Kepler capturing an extrasolar planet transit (Image credit: NASA)

Kepler's very large field of view gives it a much higher probability of detecting Earth-like planets than the Hubble Space Telescope, which, as we saw early in this book, has a field of view of only 10 square arc-minutes. Moreover, Kepler's mission is much more specialized to the detection of planetary transits, while the Hubble Space Telescope has been employed to address a wide range of scientific questions, and rarely looks continuously at just one star field. Of the approximately half-million stars in Kepler's field of view, around 150,000 stars were selected for observation. More than 90,000 are G-type stars on, or near, the main sequence. Thus, Kepler was designed to be sensitive to wavelengths of 400–865 nm, corresponding to the visible and near-infrared region of the electromagnetic radiation, where the brightness of those stars peaks. Most of the stars observed by Kepler have apparent visual magnitude between 14 and 16 but the brightest observed stars have apparent visual magnitude of 8 or lower. Most of the planet candidates were initially not expected to be confirmed due to being too faint for follow-up observations. All the selected stars are observed simultaneously, with the spacecraft measuring variations in their brightness every thirty minutes. This provides a better chance of detecting a transit.

Since Kepler must observe at least three transits to confirm that the dimming of a star was caused by a transiting planet, and since larger planets give a signal that is easier to check, scientists expected the first reported results to be larger Jupiter-size planets in tight orbits. The first of these were reported after only a few months of operation. Smaller planets, and planets farther from their sun would take longer, and discovering planets comparable to Earth were expected to take three years or longer.

Data collected by Kepler is also being used for studying variable stars of various types and performing asteroseismology (the study of sound waves in stellar atmospheres), particularly on stars showing solar-like oscillations. Once Kepler has collected and sent back the data, raw light curves are constructed. Brightness values are then adjusted to take the brightness variations owing to the rotation of the spacecraft into account. The next step is processing (folding) light curves into a more easily observable form and letting software select signals that seem potentially transit-like. At this point, any signal that shows potential transit-like features is called a threshold crossing event. These signals are individually inspected in 2 inspection rounds, with the first round taking only a few seconds per target. This inspection eliminates erroneously selected non-signals, signals caused by instrumental noise and obvious eclipsing binaries.

Threshold crossing events that pass these tests are called Kepler Objects of Interest (KOI), receive a KOI designation and are archived. KOIs are inspected more thoroughly in a process called dispositioning. Those which pass the dispositioning are called Kepler planet candidates. The KOI archive is being constantly updated, meaning that a Kepler candidate could end up in the false-positive list upon further inspection. In a similar vein, KOIs that were mistakenly classified as false positives could end up back in the candidates list.

Not all the planet candidates go through this process. For example, circumbinary planets do not show strictly periodic transits, and have to be inspected through other methods. In addition, third-party researchers use different data-processing methods, or even search planet candidates from the unprocessed light curve data. As a consequence, those planets may be missing KOI designation.

There are a few different exoplanet detection methods which help to rule out false positives by giving further proof that a candidate is a real planet. One of the methods, called Doppler spectroscopy, requires follow-up observations from ground-based telescopes. This method works well if the planet is massive or is located around a relatively bright star. While current spectrographs are insufficient for confirming planetary candidates with small masses around relatively dim stars, this method can be used to discover additional massive non-transiting planet candidates around targeted stars.

In multi-planetary systems, planets can often be confirmed through transit timing variation by looking at the time between successive transits, which may vary if planets are gravitationally perturbed by each other. This helps to confirm relatively low-mass planets even when the star is relatively distant. Transit timing variations indicate that two or more planets belong to the same planetary system. There are even cases where a non-transiting planet can also discovered in the like fashion.

Circumbinary planets show much larger transit timing variations between transits than planets gravitationally disturbed by other planets. Their transit duration times also vary significantly. Transit timing and duration variations for circumbinary planets are caused by the orbital motion of the host stars, rather than by other planets. In addition, if the planet is massive enough, it can cause slight variations of the host stars' orbital periods. Despite being harder to find circumbinary planets due to their non-periodic transits, it is much easier to confirm them, as timing patterns of transits cannot be mimicked by an eclipsing binary or a background star system.

In addition to transits, planets orbiting around their stars undergo reflected-light variations—like our Moon, they go through phases from full to new and back again. Since Kepler cannot resolve the planet from the star, it sees only the combined light, and the brightness of the host star seems to change over each orbit in a periodic manner. Although the effect is small—the photometric precision required to see a close-in giant planet is about the same as to detect an Earth-sized planet in transit across a solar-type star—Jupiter-sized planets with an orbital period of a few days or less are detectable by sensitive space telescopes such as Kepler. In the long run, this method may help find more planets than the transit method alone, because the reflected light variation with orbital phase is largely independent of the planet's orbital inclination, and does not require the planet to pass in front of the disk of the star. In addition, the phase function of a giant planet also depends on its thermal properties and atmosphere, if any. Therefore, the phase curve may constrain other planetary properties, such as the particle size distribution of the atmospheric particles.

Kepler's photometric precision is often high enough to observe a star's brightness changes caused by Doppler beaming or a star's shape deformation by a companion. These can sometimes be used to rule out hot Jupiter candidates as false positives caused by a star or a brown dwarf when these effects are too noticeable. However, there are some cases where such effects are detected even by planetary-mass companions such as in the interesting case of TrES-2b.

If a planet cannot be detected through at least one of the other detection methods, it can be confirmed by determining if a real planet is significantly larger than any false-positive scenarios combined. One of the first methods was to see if other telescopes can see the transit as well. The first planet confirmed through this method was Kepler-22b which was also observed with a Spitzer space telescope in addition to analyzing other false-positive possibilities. Such confirmation is costly, as small planets can generally be detected only with space telescopes.

In 2014, a new confirmation method called "validation by multiplicity" was announced by Kepler mission scientists. From the planets previously confirmed through various methods, it was found that planets in most planetary systems orbit in a relatively flat plane, similar to the planets found in our Solar System. This means that if a star has multiple planet candidates, it is very likely a real planetary system. Transit signals still need to meet several criteria which rule out false-positive scenarios. For instance, it has to have considerable signal-to-noise ratio, it has at least three observed transits, orbital stability of those systems have to be stable and transit curve has to have a shape that partly eclipsing binaries could not mimic the transit signal. In addition, its orbital period needs to be 1.6 days or longer to rule out common false positives caused by eclipsing binaries. Validation by multiplicity method is very efficient and allows astronomers to confirm hundreds of Kepler candidates in a relatively short amount of time.

A new validation method using a tool called PASTIS has also been developed. It enabled astronomers to confirm a planet even when only a single candidate transit event for the host star has been detected. A drawback of this tool is that it requires a relatively high signal-to-noise ratio from Kepler data, so it can mainly confirm only larger planets or planets around quiet and relatively bright stars. Currently, the analysis of Kepler candidates through this method is underway. PASTIS first proved successful in confirming the existence of the planet Kepler-420b.

Kepler's Results So Far

The Kepler observatory was in active operation from 2009 through 2013, with the first main results announced on January 4, 2010. As expected, the initial discoveries were all short-period planets. As the mission continued, additional longer-period candidates were found.

NASA held a press conference to discuss early science results of the Kepler mission on August 6, 2009. At this press conference, it was revealed that Kepler had confirmed the existence of the previously known transiting exoplanet HAT-P-7b, and was functioning well enough to discover Earth-size planets. Since Kepler's detection of planets depends on seeing very small changes in brightness, stars that vary in brightness all by themselves (variable stars) are not useful in this search. From the first few months of data, Kepler scientists have determined that about 7500 stars from the initial target list are such variable stars. These were dropped from the target list, and replaced by new candidates. On November 4, 2009, the Kepler project publicly released the light curves of the dropped stars.

The first six weeks of data revealed five previously unknown planets, all very close to their stars. Among the notable results include one of the least dense planets yet found, two low-mass white dwarfs that were initially reported as being members of a new class of stellar objects, and a well-characterized planet orbiting a binary star.

On June 15, 2010, the Kepler mission released data on all but 400 of the ~156,000 planetary target stars to the public. 706 targets from this first data set have viable exoplanet candidates, with sizes ranging from as small as the Earth to larger than Jupiter. The identity and characteristics of 306 of the 706 targets were given. The released targets included five candidate multi-planet systems. Data for the remaining 400 targets with planetary candidates was published in February 2011. Nonetheless, the Kepler results, based on the candidates in the list released in 2010, strongly suggested that most candidate planets have radii less than half that of Jupiter. The Kepler results also implied that small candidate planets with periods less than thirty days are much more common than large candidate planets with periods less than thirty days and that the ground-based discoveries were sampling the upper end of the size distribution. This contradicted traditional theory, which had suggested small and Earth-like planets would be relatively infrequent. Based on extrapolations from the Kepler data, an estimated 100 million habitable planets may exist in our galaxy! In 2010, Kepler identified two bizarre systems containing objects which are smaller and hotter than their parent stars: KOI 74 and KOI 81. These objects are probably low-mass white dwarfs produced by previous episodes of mass transfer in their systems.

Later in 2010, the Kepler team released a paper which listed data for 312 extrasolar planet candidates from 306 separate stars. Only 33.5 days of data were available for most of the candidates. However, NASA withheld data for another 400 candidates to enable members of the Kepler team to perform follow-up observations. The data for these candidates were made public on February 2, 2011 (Fig. 10.8).

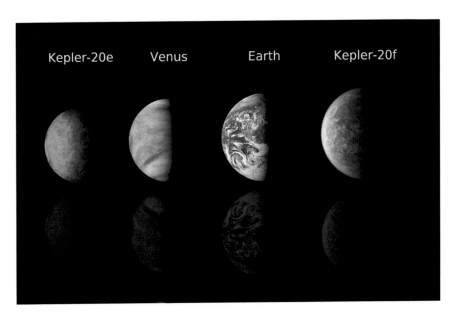

FIG. 10.8 A size comparison of the exoplanets Kepler-20e and Kepler-20f with Venus and Earth (Image credit: NASA)

On February 2, 2011, the Kepler team announced the results of analysis of observations conducted between 2 May and September 16, 2009. They found 1235 planetary candidates circling 997 host stars. Independent analysis indicated that at least 90 % of them are real planets and not false positives, 68 planets were approximately Earth-size, 288 super-Earth-sized, 662 Neptune-sized, 165 Jupiter-sized, and 19 up to twice the size of Jupiter. In contrast to previous work, roughly 74 % of the planets were found to be smaller than Neptune, most likely as a result of previous work which was more biased to finding large planets more easily than smaller ones.

That February 2, 2011 release of 1235 extrasolar planet candidates included 54 that may be in the "habitable zone", with 5 of those less than twice the size of the Earth. There were previously only two planets thought to be in the "habitable zone", so these new findings represent an enormous expansion of the potential number of "Goldilocks planets" (planets of the right temperature to support liquid water). All of the habitable zone candidates found in this survey orbit stars significantly smaller and cooler than the Sun (habitable candidates around Sun-like stars will take several additional years to accumulate the three transits required for detection).

The frequency of planet observations was highest for exoplanets two to three times the size of the Earth, and then declining in inverse proportionality to the area of the planet. From these data astronomers were able to estimate that 5.4 % of stars host Earth-sized worlds, 6.8 % host super-Earth-size candidates, a fifth host Neptune-size candidates, and just 2.55 % host Jupiter-size or larger candidates. Multi-planet systems are now believed to be common, with 17 % of the host stars have multi-candidate systems, and about 34 % of all the planets are in multiple planet systems.

By early December 2011, the Kepler team announced that they had discovered 2326 planetary candidates, of which 207 are similar in size to Earth, 680 are super-Earth-size, 1181 are Neptune-size, 203 are Jupiter-size and 55 are larger than Jupiter. Compared to the figures from February 2011, the tally of Earth-sized and super-Earth-sized planets increased by 200 and 140 %, respectively. What's more, 48 planet candidates were found in the habitable zones of surveyed stars, marking a decrease from the February figure; this was due to the more stringent criteria in use in the December data.

On December 20, 2011, the Kepler team announced the discovery of the first Earth-size exoplanets, Kepler-20e and Kepler-20f, orbiting a Sun-like star, Kepler-20. Based on Kepler's findings, astronomer Seth Shostak estimated in 2011 that "within a thousand light-years of Earth", there are "at least 30,000 habitable planets". Also based on the findings, the Kepler team has estimated that there are "at least 50 billion planets in the Milky Way", of which "at least 500 million" are in the habitable zone. In March 2011, astronomers at NASA's Jet Propulsion Laboratory (JPL) reported that about "1.4 to 2.7 %" of all Sun-like stars are expected to have earth-like planets within the habitable zones of their stars. This means there are two billion of these Earth 'analogs' in our own Milky Way galaxy alone. The JPL astronomers also noted that there are 50 billion other galaxies, potentially yielding more than one sextillion Earth analog planets, assuming that all these galaxies have similar numbers of planets to the Milky Way.

In January 2012, an international team of astronomers reported that each star in the Milky Way Galaxy may host on average…at least 1.6 planets, suggesting that over 160 billion star-bound planets may exist in our galaxy alone. Kepler also recorded distant stellar super-flares, some of which are 10,000 times more powerful than the superlative Carrington event of 1859. These superflares may have been triggered by close-orbiting Jupiter-sized planets. The Transit Timing Variation (TTV) technique, which was used to discover Kepler-9d, gained popularity for confirming exoplanet discoveries. A planet in a system with four stars was also confirmed, the first time such a system had been discovered.

As more and more data became available, the estimates were refined accordingly. By 2012, there were a total of 2321 candidates. Of these, 207 were similar in size to Earth, 680 were super-Earth-size, 1181 were Neptune-size, 203 are Jupiter-size and 55 are larger than Jupiter. Moreover, 48 planet candidates were found in the habitable zones of surveyed stars. The Kepler team estimated that 5.4 % of all stars host Earth-size planet candidates, and that 17 % of all stars have multiple planets. By this time also, two of the Earth-sized candidates, Kepler-20e and Kepler-20f, were confirmed as planets orbiting a Sun-like star, Kepler-20 (Fig. 10.9).

According to a study by Caltech astronomers published in January 2013, the Milky Way Galaxy contains at least as many

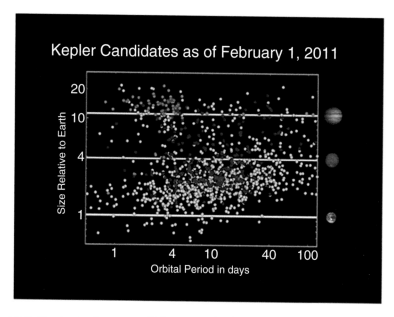

Fɪɢ. 10.9 Kepler's planet candidates as of Feb. 1, 2011 (Image credit: NASA/ Wendy Stenzel)

planets as it does stars, resulting in 100–400 billion exoplanets. The study, based on planets orbiting the star Kepler-32, suggests that planetary systems may be common around stars in our galaxy. The discovery of 461 more candidates was announced on January 7, 2013. The longer Kepler watches the more planets with long periods it can detect. Since the last Kepler catalog was released in February 2012, the number of candidates discovered in the Kepler data has increased by 20 % and now totaled 2740 potential planets orbiting 2036 stars.

Since this time, Kepler has discovered a number of notable exoplanets. A new Earth analog announced on January 7, 2013, is Kepler-69c (formerly, KOI-172.02), an Earth-like exoplanet orbiting a star similar to our Sun in the habitable zone and possibly a "prime candidate to host alien life".

In April 2013, a white dwarf was discovered bending the light of its companion red dwarf in the KOI-256 star system. At the same time, NASA also announced the discovery of three new Earth-like exoplanets—Kepler-62e, Kepler-62f, and Kepler-69c—in the habitable zones of their respective host stars, Kepler-62 and Kepler-69.

The new exoplanets, which are considered prime candidates for possessing liquid water on their surfaces, were identified using the Kepler spacecraft. A more recent analysis has shown that Kepler-69c is likely more analogous to Venus, and thus unlikely to be habitable.

Like all good things, the Kepler mission had to eventually come to an end. On May 15, 2013, NASA announced the spacecraft had been crippled by failure of a reaction wheel that keeps it pointed in the right direction. A second wheel had previously failed, and the spacecraft required three wheels (out of a total of four) to be operational for the instrument to function properly. Further testing in July and August determined that while Kepler was capable of using its damaged reaction wheels to prevent itself from entering safe mode and downlinking previously collected science data it was not capable of collecting further science data as previously configured. Scientists working on the Kepler project still had a backlog of data still to be looked at, and that more discoveries would be made in the following couple of years, despite the setback.

Although no new science data from Kepler field had been collected since the problem, an additional sixty-three candidates were announced in July 2013 based on the previously collected observations. For example, in November 2013 the second Kepler science conference was held. The discoveries included the median size of planet candidates getting smaller compared to earlier estimates, as well as the preliminary results from the discovery of a few circumbinary planets and planets in the habitable zone.

On February 13 2014, over 530 additional planet candidates were announced residing around single planet systems. Several of them were nearly Earth-sized and located in the habitable zone. This number was further increased by about 400 in June 2014.

On February 26 of the same year, scientists announced that data from Kepler had confirmed the existence of 715 new exoplanets. A new statistical method of confirmation was used called "verification by multiplicity" which is based on how many planets around multiple stars were found to be real planets. This allowed much quicker confirmation of numerous candidates which are part of multi-planetary systems. 95 % of the discovered exoplanets were smaller than Neptune and four, including Kepler-296f, were less than 2.5 times the size of Earth and were in habitable zones where surface temperatures are suitable for liquid water.

In March 2014, a study found that small planets with orbital periods of less than 1 day are usually accompanied by at least one additional planet with orbital period of 1–50 days. This study also noted that ultra-short period planets are almost always smaller than 2 Earth radii unless it is a hot Jupiter. Kepler data has also helped scientists observe and understand supernovae; measurements were collected every half hour so the light curves were especially useful for studying these types of astronomical events.

On April 17 2014, the Kepler team announced the discovery of Kepler-186f, the first nearly Earth-sized planet located in the habitable zone. This planet orbits around a red dwarf. In July 2014, the first discoveries from post-Kepler field data were reported in the form of eclipsing binaries. By September 23, NASA reported that the K2 mission had completed campaign 1, the first official set of science observations, and that campaign 2 was underway. Kepler observed KSN 2011b, a Type Ia supernova, in the process of exploding: before, during and after. Campaign 3 lasted from November 14, 2014 to February 6, 2015 and included about 17,000 new targets.

In January 2015, the number of confirmed Kepler planets exceeded 1000. At least two (Kepler-438b and Kepler-442b) of the discovered planets announced that month were likely rocky and in the habitable zone. Also in January 2015, NASA reported that five confirmed sub-earth-sized rocky exoplanets, all smaller than the planet Venus, were found orbiting the 11.2 billion year old star Kepler-444, making this star system, at 80 % of the age of the universe, the oldest yet discovered.

In April 2015, campaign 4 was reported to last between February 7, 2015, and April 24, 2015, and to include observations of nearly 16,000 target stars and two notable open star clusters, the Pleiades and Hyades in the constellation of Taurus. In May 2015, Kepler observed a newly discovered supernova, KSN 2011b (Type 1a), before, during and after explosion. Details of the pre-nova moments helped scientists to better understand dark energy. On July 24, 2015, NASA announced the discovery of Kepler-452b, a confirmed exoplanet that is near-Earth in size and found orbiting the habitable zone of a Sun-like star. The seventh Kepler planet candidate catalog was released, containing 4696 candidates, and increase of 521 candidates since the previous catalog release in January 2015.

In April 2016 NASA engineers declared a mission emergency for the Kepler probe, which had switched into emergency mode. NASA discovered the anomaly just before the agency tried to maneuver the spacecraft to point at the center of the Milky Way for a new observation campaign. Now that a mission emergency has been declared, the Kepler team has priority access to NASA's deep space telecommunications system in order to try to get the spacecraft back to normal operations. At the time of writing, it is unclear what the future holds for the spacecraft.

The Curious Case of KIC 8462852

A few astronomers used some of the world's largest telescopes to search for signals sent out by extraterrestrial intelligent civilizations. Such efforts have thus far failed, despite huge leaps in the technology needed to simultaneously search millions of frequencies. These failures, however, have done little to dampen the enthusiasm for discovering extraterrestrial intelligence. The SETI (Search for Extraterrestrial Intelligence) Institute was founded in 1984 in dedication to this scientific quest. Thanks to generous donations from Microsoft co-founder Paul Allen, the Allen Telescope Array (with its 42 antennas) has been built and dedicated to SETI observations.

Recently, astronomers have adopted a new approach to finding intelligent civilizations. Rather than looking for communication signals that may or may not be aimed at us, astronomers are attempting to detect the energy processing activities of such civilizations. This approach is based on the correlation between a civilization's level of technological advancement with the amount of energy such a civilization needs to consume to maintain such technological capability.

Russian astronomer Nikolai Kardashev defined three categories of advanced civilizations:

Kardashev I: a civilization harnessing a large fraction of its planet's incident radiation and stores of energy

Kardashev II: a civilization harnessing a large fraction of its host star's energy output

Kardashev III: a civilization harnessing a significant fraction of the energy output from its galaxy's stars

American physicist, Freeman Dyson, surmised how an intelligent civilization would proceed to harvest the energy output of a star. In the same way as putting solar panels on a roof, a civilization could surround its star or other stars with energy-capturing structures. Such structures have been labeled Dyson spheres.

Whereas previous generations of astronomers lacked the technology to detect Dyson spheres, those limitations have now been surmounted. In an Astrophysical Journal paper, four Swedish astronomers noted that Dyson spheres surrounding a large number of stars in a galaxy would change both the apparent luminosity and color of those stars as seen from Earth without changing the galaxy's gravitational potential. The team proceeded to search for those changes in the latest galaxy survey databases. In a sample of 1359 spiral galaxies (only spiral galaxies are candidates for hosting advanced life) the team failed to detect the existence of a Kardashev III level civilization.

Members of a team of 29 American and European astronomers speculated that they might have found a Kardashev II civilization. The entire team submitted a paper to the *Monthly Notices of the Royal Astronomical Society* where they reported on an observed anomalous dimming of the star KIC 8462852.5.

The dimming is not characteristic of a transiting planet or brown dwarf. It is irregular. In their paper the team considered seven possible explanations for the dimming:

Instrumental effects or data reduction artifacts

Intrinsic variability in the star's light output

Variability of KIC 8462852's companion star

Variability in light absorption by dust clouds and clumps surrounding the star

Aftermath of catastrophic collisions in an asteroid belt

Aftermath of a giant impact in the planetary system

Breakup of one or more large comet bodies that resulted in a cloud of disintegrating comets

Observations of the star and planet formation models eliminated all but the last explanation. In their summary the team concluded that a comet swarm was the most likely answer to the observed dimming.

After submitting the paper, the lead author, Tabetha Boyajian, speculated that the dimming might be caused by a Kardashev II civilization. Several interviews and Internet articles later, Boyajian and two of her colleagues garnered two weeks of time to observe on the Allen Telescope Array hoping to catch some communication signals from the possible civilization in the possible KIC 8462852 system. Alas, they found none.

Indeed, there was no real need to expend the telescope time since KIC 8462852's characteristics rule out the possibility that it could host a planet on which intelligent life exists. The spectral type of KIC 8462852 is F3V/IV, contrasted with the Sun's G2V. What is more, F3V/IV stars are nearly five times more luminous than the Sun. They burn through their nuclear fuel much more rapidly, emit much more deadly ultraviolet radiation, and manifest much more flaring activity. Any one of these distinct features would eliminate the possibility that a planet orbiting KIC 8462852 could sustain life long enough to provide the environmental conditions and bio-deposits needed for an intelligent species.

KIC 8462852's rotation period is only 0.88 days, about 30 times more rapid than the Sun's. Rapid stellar rotation is strongly correlated with high flaring activity, much too high for the survival of terrestrial animals. Also, rapid stellar rotation usually implies a stellar age much too young for the long life history an entrance of intelligent life would require.

Spectral measurements of KIC 8462852's light output revealed no significant infrared excess. The lack of excess infrared radiation contradicts the hypothesis that an intelligent civilization is using a Dyson sphere on KIC 8462852 to harvest energy. Such harvesting would involve the capturing of the star's ultraviolet and visible light to generate electricity or other forms of useful energy and the dissipation of heat (infrared radiation) resulting from such exploitation. KIC 8462852 is neither anomalously dim at ultraviolet and visible wavelengths nor is it anomalously bright at infrared wavelengths.

What KIC 8462852 is remains a mystery, but the likely answer will prove to be a new naturally occurring phenomenon of some kind. The vast universe we live in remains rich in mystery.

11. Looking to the Future: The James Webb Space Telescope

In the coming decades new space observatories are destined to revolutionize our understanding of the universe and our place within it. Of all the new space telescopes in the pipeline, it is arguably the eagerly anticipated James Webb Space Telescope that is exciting the astronomical community most (Fig. 11.1). The James Webb Space Telescope (JWST), previously known as Next Generation Space Telescope (NGST), is a space observatory currently under construction and scheduled to launch in October 2018. The JWST will offer unprecedented resolution and sensitivity from long-wavelength visible to the mid-infrared, and is a successor instrument to the Hubble Space Telescope and the Spitzer Space Telescope. The center piece of telescope houses a segmented 6.5-meter (21 feet) primary mirror and will be located near the Earth–Sun L2 point. A large sunshield will keep its mirror and four science instruments below 50 K (−220 °C; −370 °F).

JWST's capabilities will enable a broad range of investigations across the fields of astronomy and cosmology. One particular goal involves observing some of the most distant objects in the universe, beyond the reach of current ground and space based instruments. This includes the very first stars, the epoch of re-ionization, and the formation of the first galaxies. Another goal is understanding the formation of stars and planets, and accordingly JWST will be put to work imaging molecular clouds and star-forming clusters, studying the debris disks around stars, direct imaging of planets, and spectroscopic examination of planetary transits.

In gestation since 1996, the project represents an international collaboration of about 17 countries led by NASA, and with significant contributions from the European Space Agency and the Canadian Space Agency. It is named after James E. Webb, the second administrator of NASA, who played an integral role in the Apollo program.

© Springer International Publishing Switzerland 2017
N. English, *Space Telescopes*, Astronomers' Universe,
DOI 10.1007/978-3-319-27814-8_11

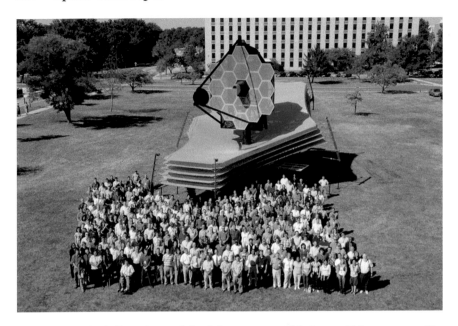

FIG. 11.1 Early full-scale model of the James Webb Space Telescope on display at NASA Goddard (2005) (Image credit: NASA)

Like the Hubble Space Telescope before it, JWST has a distinguished history of major cost overruns and delays. The first realistic budget estimates were that the observatory would cost $1.6 billion and launch in 2011. NASA has now scheduled the telescope for a 2018 launch. In 2011, the United States House of Representatives voted to terminate funding, after about $3 billion had been spent and 75 % of its hardware was in production. Funding was restored in compromise legislation with the US Senate, and spending on the program was capped at $8 billion. As of October 2015, the telescope remained on schedule and within budget, but at risk of further delays.

Evolution of a Next Generation Space Telescope

Early development work for a Hubble successor between 1989 and 1994 led to the so-called Hi-Z telescope concept, a fully baffled 4-meter aperture infrared telescope that would be paced in solar orbit at a distance of 3 AU. Here the telescope would have benefited from reduced light noise from zodiacal dust. Other early plans called for a NEXUS precursor telescope mission.

In the "faster, better, cheaper" era in the mid-1990s, NASA leaders under the auspices of Dan Goldin pushed for a lower cost space telescope. The result was the NGST concept, with an 8-meter aperture and located at L2, estimated to cost $500 million. In 1997, NASA worked with the Goddard Space Flight Center, Ball Aerospace, and TRW to conduct technical requirement and cost studies, and in 1999 selected Lockheed Martin and TRW for preliminary design concepts. In 2002, NASA awarded the $824.8 million prime contract for the NGST, now renamed the James Webb Space Telescope, to TRW. The design recommended a 6.1-meter (20 ft.) primary mirror and a launch date of 2010. Later that year, TRW was acquired by Northrop Grumman in a hostile bid and became Northrop Grumman Space Technology.

The JWST originated in 1996 as the Next Generation Space Telescope (NGST). In 2002 it was renamed after NASA's second administrator (1961–1968) James E. Webb (1906–1992), who came to fame for playing a key role in the Apollo program and establishing scientific research as a core NASA activity. The JWST is a project of NASA with international collaboration from the European Space Agency (ESA) and the Canadian Space Agency.

The telescope has an expected mass about half of Hubble's, but its primary mirror (a 6.5 meter diameter gold-coated beryllium reflector) will have a collecting area about five times larger (25 square meters vs. 4.5 square meters). The JWST is designed to conduct near-infrared astronomy, but can also image orange and red visible light, as well as the mid-infrared region, depending on the imaging instrumentation employed. The telescope will focus on the near to mid-infrared for three main reasons: high-redshift objects have their visible emissions shifted into the infrared, cold objects such as debris disks and planets emit most strongly in the infrared, and this band is difficult to study from the ground or by existing space telescopes such as the HST.

The JWST will operate near the Earth-Sun L2 Lagrange point, approximately 1,500,000 kilometers (930,000 mi) beyond the Earth. Objects near this point can orbit the Sun in synchrony with the Earth, allowing the telescope to remain at a roughly constant distance and use a single sunshield to block heat and light from the Sun and Earth. This will keep the temperature of the spacecraft below 50 K (−220 °C; −370 °F), necessary for infrared observations. Launch is scheduled for 2018 on an Ariane 5 rocket. Its nominal mission length is five years, with a goal of ten years.

NASA's Goddard Space Flight Center in Greenbelt, Maryland, is leading the management of the observatory project. The project scientist assigned to the James Webb Space Telescope is John C. Mather. Northrop Grumman Aerospace Systems serves as the primary contractor for the development and integration of the observatory. They are responsible for developing and building the spacecraft element, which includes both the spacecraft bus and sunshield. Ball Aerospace has been subcontracted to develop and build the Optical Telescope Element (OTE). Northrop Grumman's Astro Aerospace business unit has been contracted to build the Deployable Tower Assembly (DTA) which connects the OTE to the spacecraft bus and the Mid Boom Assembly (MBA) which helps to deploy the large sunshields on orbit. Goddard Space Flight Center is also responsible for providing the Integrated Science Instrument Module (ISIM).

The observatory attaches to the Ariane 5 rocket via a launch vehicle adapter ring which could be used by a future spacecraft to grapple the observatory to attempt to fix gross deployment problems. However, the telescope itself is not serviceable, and astronauts would not be able to perform tasks such as swapping instruments, as with the Hubble Telescope.

Most of the data processing on the telescope is done by conventional single-board computers. The conversion of the analog science data to digital form is performed by the custom-built SIDECAR ASIC (System for Image Digitization, Enhancement, Control And Retrieval Application Specific Integrated Circuit). It is said that the SIDECAR ASIC will include all the functions of a 20-pound (9.1 kg) instrument box in a package the size of a half-dollar, and consume only 11 milliwatts of power. Since this conversion must be done close to the detectors, on the cool side of the telescope, the low power use of this IC will be crucial for maintaining the low temperature required for optimal operation of the JWST.

The JWST's primary scientific mission has four main components: to search for light from the first stars and galaxies that formed in the universe after the Big Bang, to study the formation and evolution of galaxies, to understand the formation of stars and planetary systems and to study planetary systems and the origins of life. These goals can be accomplished more effectively by observation in near-infrared light rather than light in the visible part of the spectrum. For this reason the JWST's instruments will not measure visible or ultraviolet light like the Hubble Telescope, but will have a much greater capacity to perform infrared astronomy. The JWST will be sensitive to a range of wavelengths from 0.6 (orange light) to 28 micrometers (deep infrared radiation at about 100 K (−170 °C; −280 °F)).

The JWST will be located near the second Lagrange point (L2) of the Earth-Sun system, which is 1,500,000 kilometers (930,000 mi) from Earth, directly opposite to the Sun. Normally an object circling the Sun farther out than Earth would take longer than one year to complete its orbit, but near the L2 point the combined gravitational pull of the Earth and the Sun allow a spacecraft to orbit the Sun in the same time it takes the Earth. The telescope will circle about the L2 point in a halo orbit, which will be inclined with respect to the ecliptic, have a radius of approximately 800,000 kilometers (500,000 mi), and take about half a year to complete. Since L2 is just an equilibrium point with no gravitational pull, a halo orbit is not an orbit in the usual sense: the spacecraft is actually in orbit around the Sun, and the halo orbit can be thought of as controlled drifting to remain in the vicinity of the L2 point. This requires some station-keeping: around 2–4 m/s per year from the total budget of 150 m/s.

In order to make observations in the infrared spectrum, the JWST must be kept very cold (under 50 K (−223 °C; −370 °F)), otherwise infrared radiation from the telescope itself would swamp its instruments, greatly degrading its imaging potential. Therefore, it uses a large sunshield to block light and heat from the Sun, Earth, and Moon, and its position near the Earth-Sun L2 point keeps all three bodies on the same side of the spacecraft at all times. Its halo orbit around L2 avoids the shadow of the Earth and Moon, maintaining a constant environment for the sunshield and solar arrays. The sunshield is made of polyimide film, has membranes coated with aluminum on one side and with silicon on the other.

The sunshield is designed to be folded twelve times so it will fit within the Ariane 5 rocket's 4.57 m × 16.19 m shroud. Once deployed at the L2 point, it will unfold like an umbrella to 12.2 m × 18 m. The sunshield was hand-assembled at Man Tech (NeXolve) in Huntsville, Alabama before it was delivered to Northrop Grumman in Redondo Beach, California for testing.

The Integrated Science Instrument Module (ISIM) contains four state-of-the-art instruments and a guide camera.

The Near Infrared Camera (NIRCam) is an infrared imager which will have a spectral coverage ranging from the red end of the visible spectrum (0.6 micrometers) through the near infrared (5 micrometers). NIRCam will also serve as the observatory's wave-front sensor, which is required for wavefront sensing and control activities. NIRCam is being built by a team led by the University of Arizona, with Principal Investigator Marcia Rieke. The industrial partner is Lockheed-Martin's Advanced Technology Center located in Palo Alto, California.

The Near Infrared Spectrograph (NIRSpec) will also perform spectroscopy over the same wavelength range. It is being built by the European Space Agency at ESTEC in Noordwijk, Netherlands. The leading development team is composed of people from Astrium, Ottobrunn and Friedrichshafen, Germany, and the Goddard Space Flight Center; with Pierre Ferruit (École normale supérieure de Lyon) as NIRSpec project scientist. The NIRSpec design provides 3 observing modes: a low-resolution mode using a prism, an R~1000 multi-object mode and an R~2700 integral field unit or long-slit spectroscopy mode. Switching of the modes is done by operating a wavelength pre-selection mechanism called the Filter Wheel Assembly, and selecting a corresponding dispersive element (prism or grating) using the Grating Wheel Assembly mechanism. Both mechanisms are based on the successful ISOPHOT wheel mechanisms of the Infrared Space Observatory. The multi-object mode relies on a complex micro-shutter mechanism to allow for simultaneous observations of hundreds of individual objects anywhere in NIRSpec's field of view. The mechanisms and their optical elements are being designed, integrated and tested by Carl Zeiss Optronics of Oberkochen, Germany, under contract from Astrium.

The Mid-Infrared Instrument (MIRI) will measure the mid-infrared wavelength range from 5 to 27 micrometers. It contains both a mid-IR camera and an imaging spectrometer. MIRI is being developed in a collaborative effort between NASA and a consortium of European countries, and is led by George Rieke (University of Arizona) and Gillian Wright (UK Astronomy Technology Center, Edinburgh, part of the Science and Technology Facilities Council (STFC)). MIRI features similar wheel mechanisms as NIRSpec which are also developed and built by Carl Zeiss Optronics GmbH under contract from the Max Planck Institute for Astronomy, Heidelberg, Germany. The completed Optical Bench Assembly of MIRI was delivered to Goddard in mid-2012 for eventual integration into the ISIM.

The Fine Guidance Sensor/Near Infrared Imager and Slitless Spectrograph (FGS/NIRISS), led by the Canadian Space Agency under project scientist John Hutchings (Herzberg Institute of Astrophysics, National Research Council of Canada), is used to stabilize the line-of-sight of the observatory during science observations. Measurements by the FGS are used both to control the attitude of the spacecraft and to drive the fine steering mirror for image stabilization. The Canadian Space Agency is also providing a Near Infrared Imager and Slitless Spectrograph (NIRISS) module for astronomical imaging and spectroscopy in the 0.8 to 5 micrometer wavelength range, led by principal investigator René Doyon at the University of Montreal. Because the NIRISS is physically mounted together with the FGS, they are often referred to as a single unit, but they serve entirely different purposes, with one being a scientific instrument and the other consisting of a part of the observatory's support infrastructure.

Both the NIRCam and MIRI feature starlight-blocking coronagraphs for observation of faint targets such as extrasolar planets and circumstellar disks situated very close to bright stars. The infrared detectors for the NIRCam, NIRSpec, FGS, and NIRISS modules are being provided by Teledyne Imaging Sensors (formerly Rockwell Scientific Company).

The James Webb Space Telescope (JWST) Integrated Science Instrument Module (ISIM) and Command and Data Handling (ICDH) engineering team employ SpaceWire to send data between the science instruments and the data-handling equipment.

Northrop Grumman Aerospace Systems serves as the primary contractor for the development and integration of the observatory. They are responsible for developing and building the spacecraft element, which includes both the spacecraft bus and sunshield. Ball Aerospace has been subcontracted to develop and build the (OTE). Northrop Grumman's Astro Aerospace business unit has been contracted to build the Deployable Tower Assembly (DTA) which connects the OTE to the spacecraft bus and the Mid Boom Assembly (MBA) which helps to deploy the large sunshields when in orbit. Goddard Space Flight Center is also responsible for providing the Integrated Science Instrument Module (ISIM).

Because of its substantial distance from the Sun, the telescope itself will not be serviceable, so astronauts will not be able to perform tasks such as swapping instruments, as was the case with the Hubble Space Telescope.

Most of the data processing on the telescope will be carried out using conventional single-board computers. The conversion of the analog science data to digital form is performed by the custom-built SIDECAR ASIC (System for Image Digitization, Enhancement, Control And Retrieval Application Specific Integrated Circuit. The SIDECAR ASIC is a marvel of miniaturization that will include all the functions of a 20-pound (9.1 kg) instrument packaged inside a box the size of a half-dollar coin, and has a power consumption of just 11 milliwatts. Since this conversion must be done close to the detectors, on the cool side of the telescope, the low power use of this IC will be crucial for maintaining the low temperature required for optimal operation of the JWST when deployed in its orbit (Fig. 11.2).

JWST is the formal successor to the Hubble Space Telescope (HST), and since its primary emphasis is on infrared observation, it is also a successor to the Spitzer Space Telescope. JWST will far surpass both those telescopes however, being cable of seeing much fainter objects and much older stars and galaxies. As we have seen in a previous chapter, observing in the infrared is a key technique for achieving this, because it better penetrates obscuring dust and gas, allows observation of dim cooler objects, and because of cosmological redshift.

Prior to launch, the various components of the JWST will undergo rigorous testing to ensure that it can be deployed successfully and work flawlessly in the vacuum of space. Engineers will

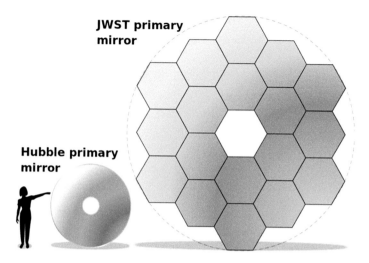

JWST primary mirror

Hubble primary mirror

FIG. 11.2 Comparative sizes of the Hubble Space Telescope (*left*) and the JWST (*right*) (Image credit: NASA)

carry out a 'center of curvature' test to precisely measure the mirror segment alignment and spacing before folding the entire optical assembly into its launch configuration. It will then be moved from the Goddard clean room into a custom made 'clean tent' for a battery of vibration and acoustic test simulating the environment it will encounter while it is carried aloft by an Arianne 5 launcher.

Toward the end of 2016, the optical assembly will be shipped to the Johnson Space Center in Houston, Texas, for further testing of all its components to ensure that the great telescope can focus starlight accurately. This will take place in a giant, thermal vacuum chamber similar to that used to test the Apollo spacecraft. Working round the clock over several months, a team of engineers will precisely measure how much light is concentrated into a point, or focused, while interferometry will measure how precisely the mirror segments can get the wave fronts from the various mirrors in the optical assembly to converge.

After that, the telescope will be shipped to Northrup Grumann in Los Angeles for integration with the rest of the telescope, known collectively as the spacecraft bus. During this final assembly process, the sunshield will be added to the package. This is the most likely component to go wrong because in launching it, project engineers have no real experience of deploying such an enormous structure in space. This will be followed by still more tests, although because of the enormous size of the assembled space telescope, there is no thermal vacuum large enough to test the whole system. At that point, project scientists and engineers will have to rely on good luck to ensure that JWST will work flawlessly.

Let's now take a closer look at the kinds of objects to be studied by JWST.

The Early Universe The more distant an object is, the younger it appears because its light has taken longer to reach us. Because the universe is expanding, as the light travels it becomes red-shifted, and these objects are therefore easier to see if viewed in the infrared. JWST's infrared capabilities are expected to let humankind see all the way to the very first galaxies forming just a few hundred million years after the Big Bang.

Seeing Through Dusty Shrouds Infrared radiation is better suited than visible rays for looking though dusty regions of space that scatter radiation in the visible spectrum. Many more stars can be seen inside dusty star forming regions in the infrared than at visible wavelengths. Observations in infrared allow the study objects such as the molecular clouds where stars are born, the circumstellar disks that give rise to planets, and the cores of active galaxies.

Studying Cool Objects Relatively cool objects (temperatures less than several thousand degrees) emit their radiation primarily in the infrared, as described by Wien's law. As a result, most objects that are cooler than stars are better studied in the infrared. This includes the clouds of the interstellar medium, the "failed stars" called brown (spectral class L) dwarfs, planets both in our own and other solar systems, icy comets and Kuiper belt objects.

The Search for Life on Other Planets

One of the main mission goals of the James Webb Space Telescope will be to study the atmospheres of exoplanets, to search for the building blocks of life elsewhere in the universe. But JWST is an infrared telescope. How is this good for studying exoplanets? One method JWST will use for studying exoplanets is the transit method, described in the last chapter, which means it will look for dimming of the light from a star as its planet passes between us and the star. Collaboration with ground-based telescopes can help astronomers measure the mass of the planets, via the radial velocity technique (i.e. measuring the stellar wobble produced by the gravitational tug of a planet), and then JWST will conduct spectroscopy of the planet's atmosphere.

JWST will also carry coronagraphs to enable direct imaging of exoplanets near bright stars. The image of an exoplanet would just be a spot, not a grand panorama, but by studying that 'spot', we can learn a great deal about it. That includes its color, differences between winter and summer, vegetation, rotation, weather and much more. How will this be done? The answer again is spectroscopy. As we have seen, spectroscopy is the science of measuring the intensity of light at different wavelengths. The graphical representations of these measurements are called spectra, and they are the key to unlocking the composition of exoplanet atmospheres.

When a planet passes in front of a star, the starlight passes through the planet's atmosphere. If, for example, the planet has sodium in its atmosphere, the spectra of the star, added to that of the planet, will have what we call an "absorption line" in the place in the spectra where would expect to see sodium. This is because different elements and molecules absorb light at characteristic energies; and this is how we know where in a spectrum we might expect to see the signature of sodium (or methane or water) if it is present.

Why is an infrared telescope key to characterizing the atmospheres of these exoplanets? The benefit of making infrared observations is that it is at infrared wavelengths that molecules in the atmospheres of exoplanets have the largest number of spectral features. The ultimate goal, of course, is to find a planet with a similar atmosphere to that of Earth. But just how likely is this?

We have yet to discover any signs of aliens, a troubling observation that has led to much speculation. One possible solution to the Great Silence is that nobody's out there. It's a conclusion that sounds impossible to believe, but there may be good scientific reasons behind it. Here's why we may be alone in the universe.

Ever since physicist Enrico Fermi famously posed the question—where is everybody?—people have been wondering why we haven't seen any signs of extraterrestrial civilizations. As Fermi pointed out, the math just doesn't add up. Our galaxy, at 13 billion years old, has been around long enough for putative aliens to explore and colonize it many times over by now (recent work shows it should take much less than a billion years, perhaps even as little as a few tens of millions of years). Clearly, we should have seen or heard from somebody by now.

This surprising observation led astronomer Michael Hart to conclude that spacefaring life in the Milky Way should be either galaxy-spanning or non-existent. But the exclusive presence of "non-existent spacefaring" aliens could be attributable to any number of things, including a reluctance to explore space, or owing to technological intractability. But it could also imply that aliens simply don't exist. Indeed, despite all the recent discoveries of potentially habitable exoplanets, along with the general feeling that our universe is primed for life, there are many reasons to suspect we're truly unique in the larger scheme of things.

As physicist Paul Davies has said, "If a planet is to be inhabited rather than merely habitable, two basic requirements must be met: the planet must first be suitable and then life must emerge on it at some stage."

Indeed, life is dependent on the presence of five critical elements, or 'metals' in the parlance of astronomers: oxygen, nitrogen, carbon, sulfur and phosphorus. These heavier elements were cooked up in nuclear reactions deep inside stars and became part of the interstellar medium only when stars reached the end of their energy-producing life. So, as time went by, the concentration of metals in the universe has gradually increased.

But here's the rub—these heavier elements only recently became sufficiently concentrated in the interstellar medium to allow life to form. Planets around older stars, therefore, are likely to be low in SPONC. Only around relatively young stars, like ours, can life emerge. So humanity would thus be among the first civilizations—perhaps the first—to arise.

But as Stephen Webb points out in his book, *Where is Everybody?*, the suggestion that chemical enrichment explains our solitude is, by itself, way too overstated—it's insufficient to completely explain the Great Silence. For example, we don't know the degree of metallicity required of a star for it to possess viable planets, and we know that the metallicity of stars vary considerable between different classes of stellar populations. Simply put, we don't know enough about this variable to make a definitive conclusion about it.

Another intriguing possibility is that our galaxy is subject to frequent gamma-ray bursts (GRBs), which we have explored earlier in the book. And by frequent we're probably talking about one every few billion years or so. As we have explored, a GRB is one of the most energetic phenomenon yet discovered in the universe. These blasts are probably caused by a hypernova—the sudden collapse of massive star to form a black hole—or the product of the collision between two neutron stars, those ultra-dense remnants of supernovae. Across the observable universe, GRBs happen at a rate of about one per day.

The ensuing blast of radiation from a hypernova has the capacity to destroy the biosphere of an Earth-like planet, instantly killing most living organisms on or near the surface (but perhaps underwater or lithoautotrophic ecosystems would survive). The gamma-rays would also instigate chemical reactions that create ozone-killing molecules powerful enough to destroy more than 90 % of a planet's ozone layer, allowing fatal ultraviolet light from the parent star to cook any complex biological molecules it strikes.

Back in 1999, James Annis of Fermilab in Illinois proposed that GRBs could cause mass extinction events on any habitable planet within a distance of 10,000 light-years from the source. To put that into perspective, the Milky Way is 100,000 light-years across and about 1000 light-years thick. Thus, a single GRB would extinguish life across a sizeable portion of the galaxy.

According to new work conducted by astronomers Tsvi Piran and Raul Jimenez, the probability that a planet could be ravaged by a GRB is strongly influenced by physical and temporal factors, that is, when and where this world exists in space and time. The closer a planet lies to the galactic core, where the density of stars is much higher than in the galactic 'suburbs', the odds of experiencing a GRB increase exponentially. Their models show that a planet near to the core has a 95 % chance of being hit by a catastrophic GRB at least once every billion years. Pulling back a bit, about half of the planetary systems in the Milky Way are close enough such that there's an 80 % chance of a GRB per billion years.

What is more, the frequency of GRBs was higher in the past owing to lower levels of metallicity in the galaxy. Metal-rich galaxies (i.e. those with significant accumulations of elements other than hydrogen and helium) feature less gamma-ray bursts. Thus, as our galaxy becomes enriched with metals, the frequency of GRBs decreases. What this means is that prior to recent times (and by recent we're talking the past 5 billion years or so), GRB extinction events were quite common. And in fact, some scientists suspect that the Earth was struck by a GRB many billions of years ago. Piran and Jimenez have surmised that these events were frequent and disbursed enough across the Milky Way to serve as constant reset buttons, sending habitable planets back to the microbial dark ages before complex life and intelligence had a chance to develop further. Intriguingly, before about 5 billion years ago, GRBs were so common that life would have struggled to maintain a presence anywhere in the cosmos.

According to Annis, "the Galaxy is currently undergoing a phase transition between an equilibrium state devoid of intelligent life to a different equilibrium state where it is full of intelligent life." Humanity, therefore, may not be alone, but one of many intelligent civilizations emerging at roughly the same time.

It's a compelling theory, but one that's still unconvincing to a growing number of skeptical scientists. Astronomer and astrobiologist, Milan M. Ćirković, a Senior Research Associate at the Astronomical Observatory of Belgrade and Assistant Professor of the Department of Physics at the University of Novi Sad in Serbia and Montenegro was at one time a big proponent of this theory, but he has since rejected this position:

He says that while the frequency of galactic GRBs is decreasing, it is not enough to explain the Great Silence. The difference between the typical timescale for the sterilizations of habitable planets and the time it would take for an advanced extraterrestrial civilization to colonize the galaxy is still way too large. Ćirković says this leaves us with two options to explain why we may be alone in the galaxy. The first is that GRBs, together with some other catastrophic process or processes (either natural or artificial), are together stifling the emergence of these civilizations. He says that together they may occur with sufficient frequency to do this, since the risks are cumulative. Alternatively, we have to seek some completely different resolution to explain the Great Silence.

One of these resolutions is the so-called Rare Earth Hypothesis—the suggestion that the parameters required to spawn a space-faring species is excruciatingly narrow. It's an idea that was put forth in 1999 by paleontologist Peter Ward and astronomer Donald Brownlee. By synthesizing the latest findings in astronomy, biology, and paleontology, the two put together a list of variables that, in their opinion, make our planet exceedingly rare in the cosmos. It is so rare, in fact, that it may explain why we may be the only ones out there (Fig. 11.3).

Fɪɢ. **11.3** Is our beautiful world unique in all the cosmos? (Image credit: NASA)

According to Ward and Brownlee, the prerequisite conditions for complex life include:

The right location in the right kind of galaxy: Galaxies have dead zones owing to varying levels of star metallicity, X-ray and gamma-ray radiation, and gravitational perturbations of planets and planetesimals by nearby stars (which can facilitate impacts of bolides). A habitable planet has to orbit at the right distance from the right type of star. Our planet is in the so-called Goldilocks Zone of our solar system where the conditions are just right for complex life to emerge. A solar system with the right arrangement of planets: Without the presence of outer gas giants, like Jupiter and Saturn, complex life may not have arisen. Interestingly, star-hugging hot Jupiter-like planets are very common.

The planet must also have a continuously stable orbit. Planets in binary systems have unstable orbits that potentially take them in and out of habitable zones. What's more, binary systems are exceptionally common in the Milky Way, accounting for at least half of all systems.

A terrestrial planet of the right size: There needs to be enough surface area, a stable atmosphere, and level of gravity that's not too heavy or light for life to flourish. A habitable planet must have continuously active plate tectonics. This process has a moderating influence on temperature excursions in the Earth's climate. A planet without plate tectonics would lack a temperature regulation mechanism—and without a stable temperature, the existence of complex life becomes increasingly unlikely. The Earth is anomalously enriched in radioactive elements like uranium and thorium that can provide the thermal energy to drive plate tectonic activity over billions of years. It is unclear at present whether other planets will be similarly enriched.

A habitable planet must have a large, stabilizing moon. Our moon puts the Earth on a stabilizing axis, allowing for seasonality, which some astrobiologists say is vital for the emergence and development of complex life. A habitable world must maintain a stable environment to allow complex life to flourish. Thus, the transition from simple cells (prokaryotes) to complex ones (eukaryotes) needs a very stable global environment.

In addition, the right kinds of organisms at the right time in cosmic history. Life has to avoid perilous environmental catastrophes, including such things as periodic bombardments of celestial objects, extreme volcanism, atmospheric factors, and as already mentioned, the increased chance of GRB extinction events.

Admittedly, this looks like a daunting list. And it just might be enough to account for the Fermi Paradox. But some astronomers think it is also possible that life is exceedingly prolific in the universe, it's just that civilization is what's rare. As Webb points out, it's not at all certain that technological species are common. As we are witness to here on Earth, complex life emerged two billion years ago, with land invertebrates arising 500 million years ago. Moreover, the origin of these animals—which sprung into existence in a narrow time window called the Cambrian explosion—has presented serious problems for evolutionary theorists. During this immense span of time, not a single one of these species developed any of these traits. Perhaps the same thing is going on elsewhere in the Galaxy, an explanation for why humanity appears to be unique. In addition, the advent of complex language, and the adoption of the scientific method are not inevitable.

There's something else that could explain our singular presence in the universe, albeit a more philosophical one. It's called the Strong Anthropic Principle (SAP), and it varies from the conventional or weak, anthropic principle in that it posits the notion that the cosmos is not just finely tuned for life—it's finely tuned for humans and humans only. SAP suggests that the conditions and parameters of the universe are so exquisitely fine-tuned for our existence as to preclude the existence of any alien civilizations. It's a highly controversial theory, but the more we learn about the universe the more it appears to be this way.

In the end though, if we truly are the first and only civilization in the cosmos, it's actually very good news. What it entails is that the future is completely open-ended and ours to shape. It also raises important philosophical questions, not least of which is the possibility that the cosmos, its exquisite fine tuning and suitability for generating life forms is the product of mind, a personal being existing beyond the space-time continuum and who may have a very particular interest in human beings.

Useful Websites on Space Telescopes

NASA's official website giving some historical and technical details of many space telescopes: https://www.nasa.gov/

An online resource chronicling the legacy of the Hubble Space Telescope: http://hubblesite.org/

A website devoted to the European Space Agency (ESA): http://www.esa.int/ESA

A data base for Kepler's exoplanet discoveries: http://kepler.nasa.gov/Mission/discoveries/

© Springer International Publishing Switzerland 2017
N. English, *Space Telescopes*, Astronomers' Universe,
DOI 10.1007/978-3-319-27814-8

Glossary

Active galaxy A galaxy that emits more energy than the light of its stars alone. Seyfert and quasars are types of active galaxies.

Apogee The point in the orbit of a satellite when it is furthest away from the Earth.

Astrometry The measurement of the positions and apparent motions of celestial objects in the sky.

Atomic number The number of protons found in the nucleus of an atom.

Black body An object which absorbs and reflects all the radiation falling on it.

Black body radiation The thermal radiation profile emitted from a black body.

Blue Stragglers Is a main-sequence star in an open or globular clusters that is more luminous and bluer than stars at the main-sequence turn-off point for the cluster. They are currently thought to have formed from interactions with other stars in the cluster giving them peculiar spectral properties.

Bow shock The space around a planet's magnetosphere where the solar wind gets deflected.

Cataclysmic variable A type of star that exhibits sudden eruptive events, where it dramatically brightens. These include nova and flare stars.

Cepheid variable A type of high mass, highly evolved star that undergoes periodic pulsations that are intrinsically related to their absolute luminosity. Cepheids were used as standard candles to establish the expansion of the universe.

Chromosphere Part of the atmosphere of the Sun located above the photosphere.

Column mass The mass of atmosphere per unit area of surface.

Compton effect The process in which a photon collides with an electron, deflecting it and causing it to lose energy in the process. It was first described by the American physicist, Arthur Compton.

© Springer International Publishing Switzerland 2017
N. English, *Space Telescopes*, Astronomers' Universe,
DOI 10.1007/978-3-319-27814-8

Corona   The outermost part of the Sun's atmosphere, extending outwards to several solar radii and reaching temperatures of up to one million Kelvin.

Coronograph   A device capable of taking images and analyzing the solar corona.

Cosmic rays   High speed subatomic particles emanating from the Sun and other stars.

Critical density   The mass density of the universe that will just cause its expansion to cease at an infinite time in the future.

Dwarf nova   A white dwarf star that has an accretion disk in its gravitational field, causing it to increase in brightness over periods as long as 1 year.

Ecliptic   The plane of the Earth's orbit around the Sun.

Electronvolt (eV)   A unit of energy equivalent to 1.6×10^{-19} joules.

Emissivity   The proportion of the radiant energy emitted by a body in comparison to its black body at the same temperature.

Filaments   Long, linear features that appear darkened against the background of the solar disk. When they occur beyond the limb of the Sun they are known as prominences.

First ionization energy   The energy required to remove the outermost electron from one mole of gaseous atoms of an element.

Flare   A small, bright emission from the Sun's surface resulting in an explosive release in energy lasting up to several minutes in duration.

Fraunhofer lines   The dark (absorption) spectral lines originally observed in the Sun by German physicist, Joseph von Fraunhofer.

Gamma rays   The most energetic form of electromagnetic radiation generally shorter than 0.01 nm emitted by the hottest astronomical bodies.

Geomagnetic index   This measures the effect of solar activity on the Earth's magnetic field, measured by continually monitoring the variations in the magnetic field near the Earth's surface.

Helioseismology   The science which investigates the interior of the Sun by studying its natural oscillation modes.

Hertzsprung-Russell (H-R) Diagram   Is a scatter graph of stars showing the relationship between the stars' absolute magnitude or luminosities versus their spectral classifications or effective temperatures. Simply put, it plots each star on a graph measuring the star's brightness against its temperature (color).

Hydrogen Alpha   The most prominent emission line in the visible region of the solar spectrum. With a wavelength of 656.3 nm (red), it is the first line of the Balmer series.

Infrared   The waveband immediately beyond the visible region of the electromagnetic spectrum from 0.8 to 300 microns.

Infrared cirrus   Large swathes of cool dust in the interstellar medium radiating at IR wavelengths (100 microns) corresponding a black body temperature of 35 K.

Ionization   The loss of one or more electrons from an atom.

Ionosphere   The region of the planetary atmosphere where atoms and molecules are ionized. The Earth's ionosphere, for example, extends from 50 to 90 kilometers above the surface.

Interferometry   Is a family of techniques in which waves, usually electromagnetic, are superimposed in order to extract information about the waves.

Interstellar Medium   The matter and magnetic fields typically found in interstellar space.

Lagrangian points   Locations in the orbital plane of two objects orbiting their mutual center of gravity where an object of small mass can remain in equilibrium.

Lyman-a   A spectral line at 121.6 nm which is produced when an electron orbiting the nucleus of a hydrogen atom in the first excited state returns to its ground state.

Magnetosphere   The region around a planet where its magnetic field has an influence.

Magnetopause   The boundary of the magnetosphere.

Main sequence   The most stable period in a star's life where hydrogen nuclei are being fused into helium in the stellar core.

Neutron star   A type of stellar remnant, originating from stars born with masses between 1.4 and 3.0 solar masses. After thermonuclear reactions cease the core collapses and protons combine with electrons to form neutrons. Neutron degeneracy pressure prevents the star from collapsing to form a black hole.

Occultation   The passage of one astronomical object in front of another resulting in one object becoming invisible for a finite period of time.

Perigee   The point in the orbit of an earth orbiting satellite where it is nearest the Earth.

Photon   A quantum mechanical description of a light 'particles' delivering fixed quantities of energy directly proportional to their frequency.

Photosphere   The visible surface of the Sun, which has a temperature of 6000 K.

Plages   Bright areas seen in the chromosphere associated with sunspots.

Plasma   An ionized gas containing electrons and ionized atoms. Plasma is the most abundant state of matter in the cosmos.

Prominences   Solar filaments seen on the edge of the solar disk.

Protostar   A newly formed star that has just begun thermonuclear fusion in its core.

Pulsar   A rapidly rotating neutron star which emits a strongly focused beam of radio waves

Quasar   A very bright object located at extragalactic distances with a high red shift. Its great luminosity is derived from a very active galactic nucleus.

Redshift   A shift in the spectral lines towards the red end of the spectrum, the extent of which reveals the recessional velocity of the object.

It is opposite to blueshift, which measures velocity of approach along our line of sight.

Scintillation counter   A detector consisting of an array of crystals that emit flashes of light when a gamma ray hits it. Each flash is amplified using a photomultiplier.

Seyfert galaxy   A type of active galaxy, with a brilliant point-like nucleus and inconspicuous spiral arms, exhibiting a very broad emission spectrum.

Spherical aberration   An optical aberration where the rays of light from the edge of a lens or mirror come to focus at a different location to those on the optical axis.

Spicules   Small spike like appendages seen at the top of the solar chromosphere.

Standard candle   A celestial object e.g. Cepheid variables or supernovae that can be used to measure vast astronomical distances by virtue of their intrinsic physical properties.

Supernova   The explosion of a massive star at the end of its life (Type II), or white dwarf (Type I) in a binary star system that has accreted mass from its companion.

Synchrotron radiation   Electromagnetic radiation (usually radio waves) produces when electrons spiral down a magnetic field.

Tomography   The astrophysical technique of producing cross sectional images of stars.

Transition region   The region of the Sun located between the corona and the chromosphere where the temperature soars from about 20,000 K to over 1,000,000 K.

Ultraviolet   The waveband beyond the violet end of the visible spectrum from 10 to 380 nm.

Van Allen Belts   A pair of ring-shaped belts consisting of electrically charged particles trapped. The inner belt is found between 2000 and 5000 kilometers above the earth's surface, while the outer belt is found about 13,000 and 20,000 kilometers from the Earth.

Visible light   Radiation detectable by the normal human eye from about 390 nm to 700 nm.

Wolf-Rayet stars   A type of young, hot, O-spectral class stars displaying a spectrum with strong, broad emission lines.

White dwarf   A type of stellar remnant representing the core of solar mass stars that have shed their outer atmospheres to interstellar space. Consisting of electrons and protons, they are held up against gravitational collapse via electron degeneracy pressure.

X-rays   A type of high energy electromagnetic radiation emitted by a variety of celestial objects covering a waveband between 0.01 to 9 nm.

Zeeman Effect   The phenomenon whereby a spectral line is split into two in the presence of a strong magnetic field.

Bibliography

D. Baker, *Spaceflight and Rocketry: A Chronology* (Facts on File, New York, 1996)

E.J. Chaisson, *The Hubble Wars: Astrophysics Meets Astropolitics in the Two Billion Dollar Struggle Over the Hubble Space Telescope* (Harvard University Press, Cambridge, 1998)

J.K. Davies, *Astronomy from Space, The Design and Operation of Orbiting Observatories* (Wiley, Chichester, 1997)

M. Elvis (ed.), *Imaging X-ray Astronomy, A Decade of Einstein Observatory Achievements* (Cambridge University Press, Cambridge, 1990)

J. Gribbin, *Alone in the Universe* (Wiley, Hoboken, 2011)

W. Hardwood, *NASA's Next Generation Space Telescope Takes Shape* (Astronomy Now, 2016), pp. 54–58

R. Hirsh, *Glimpsing an Invisible Universe: The Emergence of X-ray Astronomy* (Cambridge University Press, Cambridge, 1983)

K. Huffbauer, *Exploring the Sun: Solar Science Since Galileo* (John Hopkins University Press, Baltimore, 1991)

D. Leverington, *New Cosmic Horizons, Space Astronomy from the V2 to the Hubble Space Telescope* (Cambridge University Press, Cambridge, 2000)

C.C. Petersen, J.C. Brandt, *Hubble Vision; Astronomy with the Hubble Space Telescope* (Cambridge University Press, Cambridge, 1995)

K.J.H. Philips, *Guide to the Sun* (Cambridge University Press, Cambridge, 1995)

G. Smoot, K. Davidson, *Wrinkles in Time: The Imprint of Creation* (Abacus, London, 1995)

J.N. Tatarewwicz, *Space Technology and Planetary Astronomy* (Indiana University Press, Bloomington, 1990)

© Springer International Publishing Switzerland 2017
N. English, *Space Telescopes,* Astronomers' Universe,
DOI 10.1007/978-3-319-27814-8

F. Taylor, *Exploring the Planets: A Memoir* (Oxford University Press, Oxford, 2016)

P. Ward, D. Brownlee, *Rare Earth: Why Complex Life is Uncommon in the Universe* (Copernicus Books, New York, 2003)

S. Webb, *If the Universe is Teeming with Aliens - Where is Everybody?: Fifty Solutions to Fermi's Paradox and the Problem of Extraterrestrial Life* (Copernicus Books, New York, 2002)

M. Zeilik, S.M. Gregory, *Introductory Astronomy and Astrophysics* (BrooksCole, Pacific Grove, 1997)

Index

© Springer International Publishing Switzerland 2017
N. English, *Space Telescopes*, Astronomers' Universe,
DOI 10.1007/978-3-319-27814-8

Printed in the United States
By Bookmasters